COMMUNITY HEALTH
A Systems Approach

Unless Recalled Earlier

COMMUNITY HEALTH
A Systems Approach

CARRIE JO BRADEN, M.S.

Assistant Professor
Assistant to the Dean
Mississippi College
School of Nursing
Clinton, Mississippi

NANCY L. HERBAN, M.S.Hy.

Assistant Professor
Director of Graduate Program
Mississippi University for Women
School of Nursing
Columbus, Mississippi

APPLETON-CENTURY-CROFTS/New York

Library of Congress Cataloging in Publication Data

Braden, Carrie Jo, 1944-
    Community health: a systems approach.

    Bibliography: p.
    Includes index.
    1. Community health nursing. 2. System analy-
sis. I. Herban, Nancy L., joint author. II. Title.
[DNLM: 1. Community health services—Nursing
texts. 2. Systems analysis—Nursing texts. WY106
B798c]
RT98.B7   362.1   76-20776
ISBN 0-8385-1189-8

Copyright © 1976 by APPLETON-CENTURY-CROFTS
A Publishing Division of Prentice-Hall, Inc.

76 77 78 79 80 / 10 9 8 7 6 5 4 3 2 1

Prentice-Hall International, Inc., London
Prentice-Hall of Australia, Pty. Ltd., Sydney
Prentice-Hall of India Private Limited, New Delhi
Prentice-Hall of Japan, Inc., Tokyo
Prentice-Hall of Southeast Asia (Pte.) Ltd., Singapore

PRINTED IN THE UNITED STATES OF AMERICA

cover design: Karen Robbins

DEDICATION

To friends, sons, and lovers

ACKNOWLEDGMENTS

We would like to acknowledge those without whom this volume would not have been written—the students and faculty of Indiana University School of Nursing (with whom we were associated), who gave us input, support, and feedback as the models were developed, tested, and evaluated; to Dr. Shirley Burd, who saw potential in the project and introduced us to our editor, Mr. Charles Bollinger; to Mr. Bollinger, who has provided us with moral support as we struggled toward the finishing date; and to our families and others who were patient, understanding, and supportive through the developmental stages of the project.

Carrie Jo Braden
Nancy L. Herban

ACKNOWLEDGMENTS

CONTENTS

INTRODUCTION

Our life is a faint tracing on the surface of mystery, like
the idle, curved tunnels of leaf miners on the face of a
leaf. We must somehow take a wider view, look at the
whole landscape, really see it, and describe what's going
on here.

ANNIE DILLARD, 1974[1]
Pilgrim at Tinker Creek

Each year increasing emphasis is placed on health and a health care delivery system that is both community focused and community based. This
community emphasis places greater responsibility upon educational institutions to provide content and related learning experiences that will prepare
their graduates to assume leadership roles in this ever-changing world.

Health is a community affair. Inherent in the philosophy and definition
of community health and community health nursing is the belief that man-
environment relationships are strong determinants of health. However, the
search for a method of presenting content that could provide a rational,
holistic approach to the study of the community has been difficult. The
challenge, accepted by us, was to find a way to assist the student, accustomed to the insulated hospital setting, to "take a wider view, look at the
whole landscape, really see it, and describe what's going on here." Only
through expansion of awareness and knowledge can the processes for management of health be applied to the community. Taking the "wider view" will
enable the practitioner, regardless of discipline, to make comprehensive
assessments, to plan appropriately, to implement change effectively, and to
evaluate progress as accountable providers and consumers of health care
services.

Our search led us to consider and apply general systems theory* as the framework for the study of the community. Anatol Rapoport has stated: "Human social aggregates (families, institutions, communities, nations) exhibit all the features of organized systems."[3] Charles Loomis[2] has suggested specific common elements and processes that are common to all systems and that could be applied to describe and compare systems. We believe the systems theory approach provides a theoretic framework that is practical, applicable, and timely. This is especially true as professionals increase their interdisciplinary approach to health care. The systems theory approach provides a methodology that utilizes a common language which can enhance interdisciplinary communication. The conceptual models and visualizations will assist the student in the comprehension, synthesization, and application of content. To quote Rapoport, ". . .the system approach to the study of man can be appreciated as an effort to restore meaning (in terms of intuitively grasped wholes) while adhering to the principles of disciplined generalizations and rigorous deduction. It is in short an attempt to make the study of man both scientific and meaningful."[3]

This book has been divided into three sections: Section I: Community: A Systems Approach provides the basis or introduction to the other sections. Chapter 1 considers general systems theory, including background material, basic definitions, and discussion of the elements and processes common to all systems. A theoretic model will be presented. Chapter 2 considers a limited application of the theoretic model to the family and community and an in-depth discussion of the model as applied to the health care system. Such items as factors that affect boundary maintenance, organization, structure, etc., will be included. Section II: Community Health Decision-Making comprises Chapter 3: Assessment; Chapter 4: Planning and Implementation; and Chapter 5: Evaluation. Each of these chapters will include a discussion of some tools and resources that are available and that can be used to accomplish the task at hand. Section III: The Future will contain speculations and hypotheses regarding the community as well as the future of health care and delivery of health services.

This book is not designed to provide solutions to the multiple problems of the community, but rather is presented as one approach to the study of the community. We believe that the systems approach is logical, not only for nursing, but for other disciplines which are concerned with the study of the community and health. We believe that this book provides a basic format and that the level of sophistication of its presentation will be dependent upon each individual faculty using it and the level at which such content is placed within the curriculum pattern.

Appendix A provides a glossary of terms used throughout.

REFERENCES

1. Dillard A: Pilgrim at Tinker Creek. New York, Harper, 1974
2. Lommis C: Social Systems—Essays on their Persistence and Change. Princeton, N.J., Van Nostrand, 1960
3. Rapoport A: Foreword. In Buckley W (ed): Modern Systems Research for the Behavioral Scientist. Chicago, Aldine, 1968

Section I
COMMUNITY:
A SYSTEMS APPROACH

INTRODUCTION

Physical science and social science professionals have been involved in a game of one-upmanship. Sciences in which the analytic method worked well and in which results of studies could be stated in measurable terms were described glowingly as being "pure" sciences. These sciences easily fit the criteria for a profession. Sciences in which analytic approaches did not work and in which exact predictable results did not occur were described as "sloppy," and their professional status has often been questioned based on the imprecision of their knowledge base. The most common retaliation made by workers in these latter sciences was that living processes are not governed by the same laws as nonliving processes. Living processes have an esoteric aura that have given social scientists a certain immunity to professional "digs" from those involved with "mere" things. Thus, the game of who really is dealing with valuable scientific material continued without resolution.

Health care professionals have often been caught in this philosophic bind. Health care disciplines are concerned with the chemical-biophysical components and psychologic-sociologic aspects that influence an individual's health status. Their professional knowledge base is drawn from the physical and social sciences, so professional loyalties are often split. Relationships among health care professions are also splintered by the fact that some disciplines adhere more closely to analytic theory than to vitalist theory. The traditional medical model approach to health care is basically analytic in origin. It relies heavily on the paradigm of scientific assertion. "If so . . . then so." (All things being equal, when a specific bacterial agent is introduced into a human organism, certain specific symptomatology will occur and certain specific medical interventions will interrupt this symptomatology, thus

2

returning the human organism to a well state.) This approach does not include input from the social sciences. It also antagonizes health professionals who through their study of environmental influences have come to realize that living organisms, particularly human organisms, cannot be studied in a completely controlled setting. Such disciplines maintain that man cannot be studied in a causal relation free from disturbances by other factors. The resulting professional backbiting has not facilitated movement toward improved health care.

Such dichotomies need not exist between the physical and social sciences. Research to date has done much to disprove the idea that living processes function by different rules than nonliving processes. Theoretically then, human behavior (physical and social) could be subjected to scientific analysis. But the scope of analytic methods does not expand sufficiently to include "concepts which embody irreducible wholes in place of physically measurable variables."[1] There is a theoretic point of view which is applicable to both social and physical sciences and which meets the need for a methodical study approach. This viewpoint has the added advantage of presenting itself in a common language that is shared across all professional discipline lines. General systems theory studies "wholes" which function "holistically" because of the interdependence of their parts. It is derived directly from analytic thought so is compatible with it. General systems theory can be applied to physical systems which "are usually simple enough to be understood entirely in terms of the interrelations of their parts"[1] and to social systems in which processes are too complex to yield to the analytic method.

In summary, "the system approach to the study of man can be appreciated as an effort to restore meaning (in terms of intuitively grasped understanding of wholes) while adhering to the principles of disciplined generalization and rigorous deduction. It is an attempt to make the study of man both scientific and meaningful."[1]

REFERENCE

1. Rapoport, A: Foreword. In Buckley W (ed): Modern Systems Research for the Behavioral Scientist. Chicago, Aldine, 1968

1

GENERAL SYSTEMS THEORY

... systems theory, the most powerful tool we have today
for effecting the unification of scientific knowledge, and
the utilization of that knowledge for humanistic ends.

ERVIN LASZLO, 1972
The Relevance of General Systems Theory[8]

A system is a whole which functions as a whole because of the
interdependence of its parts. Systems theory studies the patterns of
interactions among the system's parts, derives from these interaction patterns
"laws" of their functioning, and organizes classes of systems from wholes
whose patterns of interactions are governed by the same "laws." Use of a
conceptual model is one way systems theorists "picture" interaction patterns
for study.

One of the basic systems concepts has to do with defining the
environment of the system. Conceptually, a picture of this definition can take
a variety of forms. The mathematician may use an equation and mathematical
operations

$$E_s f \left\{ C \xleftrightarrow{\ R\ } A \right\} .$$

A psychologist might draw his conceptual picture in narrative form: the
environment of a system is determined by the system components, their

5

attributes, and the relationship between these attributes and components. A physicist might use a diagram flow chart such as illustrated in Figure 1. This last approach of picturing systems interaction patterns is the method most consistently used in the presentation of systems theory in this volume.

Kenneth E. Boulding, in his article, "General Systems Theory: The Skeleton of Science,"[4] describes a hierarchy of system levels which helps to categorize the complexities of systems theory concepts.

He discusses in detail nine such levels, beginning with static structures, continuing with the arrangement of atoms in a crystal, and ending with transcendental systems or the ultimates, absolutes, and inescapable unknowables which also exhibit systematic structure and relationship. These nine levels can more generally be divided into the following three classifications:

1. Physical or mechanical systems—the basis of knowledge in the physical sciences such as physics and astronomy.
2. Biologic systems—the interest of biologists, botanists, and zoologists.
3. Human and social systems—the concern of the social sciences as well as the arts, humanities, and religion.

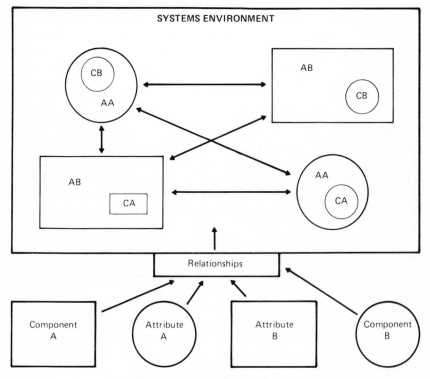

FIG. 1: Flow chart of a system.

The degree of complexity increases with each succeeding classification level. Physical and mechanical systems are basically structural in nature. Biologic systems add process elements to the structural framework which multiplies the interaction pattern possibilities. Human and social systems compound the intricacy of relationship patterns, as many of the structural and process elements in social systems are abstract concepts and ideas rather than concrete factors. The term *equifinality* describes this progressive complexity of interaction patterns found in systems. Predictive studies in complex systems are difficult, because the final product of a system's interactions may be reached from different initial conditions and in different ways. Systems that seem to have the same initial condition and the same structural/process elements can end with entirely different products. It is not unusual, then, that most research efforts have been centered in the simple system classifications of physical and biologic sciences. The unifying language of predictive research has been mathematics. But because of the equifinality effect, mathematics' usefulness in complicated system functioning has been limited. The traditional one-cause-equals-one-effect relationship does not agree with general systems theory concepts.[9]

Another systems theory term describing relationship among components, is nonsummativity. *Nonsummativity* defines the degree of interrelatedness among system parts. The higher the degree of nonsummativity the greater the interdependence of components. A high degree of nonsummativity is illustrated when a change in one part of the system effects change in all other parts and in the total system. Systems in which the degree of nonsummativity is decreasing are in the process of becoming nonfunctional. A system does not work without some interdependence among components. Components need to have some degree of nonsummativity (interrelatedness) in order for a system to function. Consider the short story, "The Red-Headed League," by Sir Arthur Conan Doyle, featuring Sherlock Holmes as an example.[5] In this story, red-headed men who are sound in body and mind and above the age of twenty-one years are brought to one location by the "stimulus" of a newspaper ad in *The Morning Chronicle.* The "components," red-headed men, seem to be "interrelated" in that they have common physical characteristics and they have come into close contact with one another. In reality, however, these components have a low degree of "nonsummativity," as they do not interact or interrelate with one another. They merely stand in line for a time and then disperse when the announcement comes that the requirements as set forth in the ad have been filled. On the other hand, later in this same story some components (ie, a red-headed pawnbroker, a physician, a private investigator, a policeman, and a bank official) do interact and form a "system" which eventually fulfills its purpose in catching a notorious crook.

Because the study of systems requires looking at many variables, it can become difficult to determine where one system ends and another begins. In order to facilitate the identification of any given system's boundary, the *focal unit* of approach must be identified. In the fields of physical and biologic science, systems are most often studied in a very small focal unit; study of a single cell, study of molecular structure, study of a single chemical interaction would be examples of focal units in the physical and biologic sciences. Social sciences also study small system units—interaction between two people, a family, a small group. When the focal unit is on a small scale, it can be called study of a *microsystem*. As the focal unit becomes more complex and incorporates several microsystems (subsystems), the focus of study is said to be at the *mezzolevel*. Study of the nervous system, study of socially developed programs and policies that affect large groups are studies of mezzosystems. Studies of systems which incorporate large and/or complex interrelationships such as the anatomic and physiologic systems which interrelate to make up a functional human being are focused at the *macrolevel*. Social sciences which study systems affecting large geographic and scattered populations also involve macrolevel approaches. Current systems analysis tools are much more highly defined at micro- and mezzolevels than at the macrolevel. Physical and biologic sciences have made more advances in systems study at all focal levels, while social sciences are making the most active use of general systems theory at the microlevel.

There is one other concept necessary to an understanding of system function. Systems can operate either as *open* or *closed* units. Input-output channels for exchange of data with the environment outside the system are necessary for an open-system operation. All living organisms are open systems. Their *feedback loops* are channels through which both positive and negative feedback enter and leave the system. *Positive feedback* comprises data that lead to change; *negative feedback* comprises data that maintain stability. Closed systems exist in the field of physical science as classic chemical or physical models in which no exchange occurs with the environment. Social scientists use the term *closed system* to refer to systems that do not accept or use feedback readily. Such social systems are usually rigidly organized and slow to change with the times.

In summary, elements necessary for system function are both structural and transactional. Their degree of interaction is stated in terms of wholeness or nonsummativity, and the increasing complexity of these interrelations during the system process is called equifinality. Feedback loops and the acceptance of both positive and negative feedback are essential for open-system operation. Closed systems characteristically make limited use of feedback and are slow to change. It is necessary to determine the focal unit before a system can be analyzed, as increasing levels of complexity lead to systems functioning at micro-, mezzo-, and macrolevels. Study methods for micro- and mezzolevel systems and most physical science systems are better developed than those methods for study of social systems and macrolevel systems.

INTERNAL SYSTEM OPERATION

In an attempt to bridge the gap between the physical, biologic, and social sciences, a conceptual model which pictures systems operation is presented in the following text and in Figures 2 to 10. This model can be used to trace energy flow through either a physical system or social system. It is also applicable at micro-, mezzo-, and macrolevels, depending on the focal unit of application. Figure 2 shows the conceptual model in its totality.

The model presented in this chapter is an analogy, so in application it can tolerate some inconsistencies not in accord with the real phenomenon. For example, the stop-action approach taken here never truly occurs in system function, as all processes and elements are in constant interaction all the time. This model is only a device to help explain relationships of variables under study; it is not a theory within itself. General Systems Theory also is a conceptual scheme that explains relationships of variables under study. But as a theory, all applications must be compatible with the real phenomenon. Any facts that are not in accordance with the applied theory would invalidate the theory. General Systems Theory has been validated enough times over the last fifty years to prove its truthfulness in application. The model presented here needs to be judged by its usefulness in explaining how a system works. This model is not a theory; it is a tool to help study a theory.

Because systems are composed of dynamic ongoing interactions, any "picture" of a system is merely a stop-action representation. A system must have stability over time and space but a conceptual model that would attempt to show interaction over time and space would look like a series of stop-action shots found on a motion picture film. The following models are all based on the premise that this stop-action "picture" exists only in a split second of time and transitional position in space, and that such a picture does have stability over time and space.

All systems need energy in order to function. Inert energy depicted in the conceptual model as an energy pool (Fig. 2) was selected as the structural element most basic to all system functions. An available energy source (structure) was the philosophic element that begins this diagrammatic representation of system function. According to the rules of concept development, a process element should follow. This is pictured in a simple language and called *energy transformation* (Fig. 2, step 1). In order for this energy to become transformed, a stimulus must hit the energy pool and shake

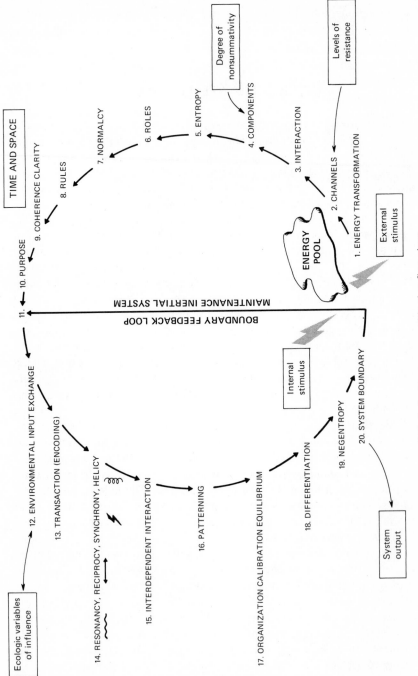

FIG. 2: Systems energy utilization flow chart.

it up. The source of this stimulus could be from outside the system or from feedback fed through the energy pool from its own feedback loop mechanism. This stimulus (whatever its source—internal or external) is the *major* encoding unit for the system's initial functional status (Fig. 3).

Transformed energy or activated energy radiates out into the system through *channels* (Fig. 2, step 2). Specific names for these channels depend upon the nature of the system being studied. Receptor cells have axons which serve as channels for cellular input. Electric conduits are channels for electric charges. What is consistent about all these kinds of channels is that each channel has its own characteristics that influence the flow of energy on through the system. All channels present a certain degree of resistance to incoming energy as well as a certain degree of receptiveness (Fig. 4). Channel sensitivity to energy can be measured in some systems—physical and biologic systems theory have developed several tools for such measurement. The importance of this attribute is that if the level of channel resistance is high, the energy input will be absorbed before interaction can occur and no further system process can occur. Channels also have an upper limit of energy acceptance. This upper limit can be called channel capacity. Once this capacity has been reached, the channel will accept no additional energy input. Some systems have built-in circuit breakers to diffuse excess energy at this point and prevent channel damage. Other systems with no such regulators can break down through channel overload or inherent channel damage.

Because of these channel characteristics and resultant accommodations to energy flow, energy *interaction* (Fig. 2, step 3) patterns develop. These patterns evolve from the energy changes incurred through contact with the channels. Ervin Laszlo, in his book, *The Systems View of the World,*[7] has described this process as follows:

> *Imagine a universe made up not of things in space and in time, but of patterned flows extending throughout its reaches. What flows is a mysterious, non-individualized something we call energy. It flows along*

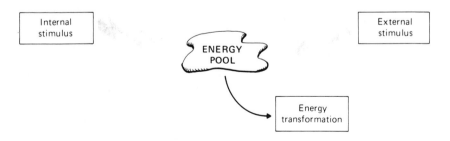

FIG. 3: Conceptual model: energy transformation.

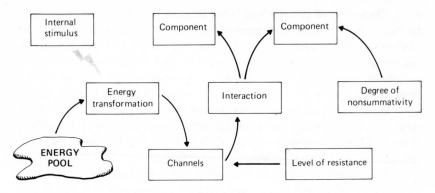

FIG. 4: Regulation boxes: beginning internal systems functions.

pathways structured by the metric of integral space-time. It flows smoothly, without crinks or wrinkles, over vast stretches of this cosmic matrix, and it becomes contorted in some regions. In these regions there are disturbances along the flows induced by the presence of electromagnetic forces. Some of the flows tie themselves into knots and twist into a relatively stable pattern. Now there is something there— something enduring—whereas before there was but a transitory plan. . .[11]

Laszlo continued to discuss "particles of matter" which are pulled into system organization by energy interaction patterns. *Components* (Fig. 2, step 4) are the knots found in this conceptual model's energy flow (Fig. 4). Components swept along with this tidal force will not contribute toward system process unless there is some degree of interdependence among them. The concept of nonsummativity is operational at this point of systems process.

Even components that are interdependent and thus functional for continued systems process have little stability at this point of systems activity. The Second Law of Thermodynamics, a classic physics law, accounts for this instability or *entropy* (Fig. 2, step 5). This law defines entropy as a measurable physical quality. It is based on the statistical tendency of matter to go over into disorder.

This disordered state is characterized by decrease in usable energy.[11] Components absorb a great deal of the usable energy by acting as a drag on the free flow. Since at this level of systems process there is no other source of energy input, it would be possible for increased entropy to occur. A closed system would not make full use of the feedback loop and thus have little or no access to outside energy sources (feedback loop, Fig. 2, steps 11 to 20). Without the energy inputs which occur at the feedback loop segment, the existing energy level needed to run the system would be depleted quickly. Increased entropy leads to death of a system. Any system has the potential for ending itself at this point if the energy available is not sufficient to carry functions through to the feedback loop. It is like swinging a heavy object at the end of a string. If the object is swung in a vertical circle and allowed to pivot only under its own strength, it will slow as it reaches the top of the loop. With each succeeding circle, the object will continually slow until it has no energy left to make the uphill part of the oscillation (Fig. 5).

Entropy need not end every system. In the conceptual model shown in Figure 2, one of the assumptions made is that at each step there is sufficient energy force to continue on to the next. Discussion of each of these steps will include possibilities for breakdowns as well as progress to the next step. One other concept should be understood at this point. Although the Second Principle of Thermodynamics is operational in all systems (open and closed), it has more immediate effect in closed systems. The Second Principle does not, however, operate under a single specific time limit. Decay or death of a system may extend over a long period of time. In the case of open systems, if input through the feedback loop causes sufficient internal operational changes, the system itself is then different from its initial state and the effect of entropy begins again in a new time frame. If changes keep ahead of entropy's death sentence, a system can remain functional for an indefinite time limit.

When entropy fails to dissipate the energy available for systems function, the energy is then free to move onto another systems process. The Principle of Organizing Energy Level, as stated by Gordon Allport,[2] assumes control of the system's existing energy level. This principle states that "there must be

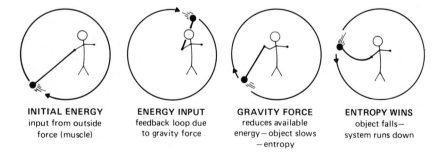

INITIAL ENERGY	ENERGY INPUT	GRAVITY FORCE	ENTROPY WINS
input from outside force (muscle)	feedback loop due to gravity force	reduces available energy — object slows — entropy	object falls— system runs down

FIG. 5: Entropy.

motives to consume one's available energies; and if existing motives do not suffice, new ones will develop."[2] Allport has made use of a well-known physics theory that energy exists as either "bound energy" (the bound energy is the entropy times the absolute temperature) or unavailable heat or "free energy," which has the potential for being used productively. Free energy undirected will soon return to a lower level of integration. For a system to continue to function, then, it must at this point make some use of the "free" energy and it must be a use which requires a more complex arrangement of the components or the energy will be lost to system operational needs.

The component arrangements have been designated, in the conceptual model, as *roles* (Fig. 2, step 6). Roles are an avenue along which nature can create order out of disorder. Roles are only theoretic constructs and cannot be visualized directly in any systems operations. Observation of component behavior is the only way to see a role. It is necessary to make a dual observation—not only must behavior be noted but stimuli must also be monitored. Over a period of time, it can be seen that any one component will change little or not at all in response to certain classes of stimuli but to other classes of stimuli it may react with behaviors that have a high degree of interchangeability. From such observations, roles in response to equivalent stimuli can be *inferred*.

Laszlo[7] attempts to clarify the process of role definition through an elaboration of his knots in space analogy.

> *Let us suppose that there are a vast number of such knots tied across the reaches of space-time, and that these knots are at uneven distances from one another. They form not isolated units but part of a continuum. Their primary mode of communication is attraction and repulsion depending on the distance separating them from one another.*

Attraction and/or repulsion seems to be the most basic behavior that contributes to any component's acceptance or rejection of a role. The conceptual model shows that each component has developed a certain degree of individualization because of the effect of a specific energy stimuli, and because of this individualization each component reacts differently to stimuli. These stimuli, or vector forces, move components toward regions of positive acceptance or away from areas of negative acceptance. Research in biophysics has demonstrated that even the most simple systems react to both positive and negative stimuli.[12] This most elementary response is the beginning of the role differentiation. Von Bertalanffy in his book, *General Systems Theory,*[15] talks of this differentiation as being a change from a more general and homogeneous condition to a more specialized and heterogeneous state. The developmental continuum of a system moves from relative globalness and lack of differentiation to a state of increasing articulation and hierarchic order. Laszlo[7] expands further on this process of role definition.

> *... regardless of whether we are talking about physical systems, species of living organisms, or social systems, we find that those which are likely to be around are the ones which are hierarchically organized. In fact, there are no records of any others; they are too vulnerable to develop to any appreciable extent.*

As systems considered become more complex, there is some argument whether or not the stimulus response phenomenon is sufficient to explain the manifold implications of role development. The wide variation found in behavior of highly organized systems seems to be too extensive to be explained by a stimulus response reaction. Such criticisms do not take into account the levels of interdependence found in system functions. Laszlo explains this interdependence as "... freedom bounded by the limits of compatibility with the dynamic structure of the whole."[7]

Once components have tried on a series of roles and found those which are functional, a period of *normalcy* (Fig. 2, Step 7) follows. Allport[1] describes this process in a series of quasimechanical principles. The Law of Effect states that action which is rewarded will be repeated. The Principle of Mastery states that whatever increases or enhances will be kept, and the Principle of Perseverative Functional Autonomy states, "a mechanism set in action because of one motive will at least for a time feed itself."[1]

Out of this normalcy, *rules* (Fig. 2, step 8) of function form. Some systems operate on very formal concrete redundant laws; others form more flexible rules. The energy of the system is becoming more usable as it is guided from one system process to another. Such usability promotes *coherence* (clarity of purpose; Fig. 2, step 9) among the components and their attributes. Out of this clarity, a functional *purpose* for the system's existence is formulated. What was once free energy serving no specific purpose and what were once components idling in space are now together a part of a system which operates in time and space for a specific purpose.

To this point we have traced the flow of energy through internal system operations. Steps 1 through 10 on the conceptual model (Fig. 2) depict this energy flow. These steps are the body of systems organization. The remainder of this text material and the conceptual model deal with feedback loop operations. Systems can function without making use of this feedback loop segment.

Allport[1] describes what happens to systems which do not make use of the feedback loop.

A closed system is defined as one that admits no matter from outside itself and is therefore subject to entropy according to the Second Law of Thermodynamics. While some outside energies, such as change in temperature and wind (time and space influences) may play upon a closed system—it (the system) has no restorative properties and no transactions with its environment so that like a decaying bridge it sinks into thermodynamic equilibrium.

He goes on to describe four criteria found in all definitions of open systems. These are

1. Intake and output of both matter and energy.
2. Achievement and maintenance of steady (homeostatic) states so that intrusion of outer energy will not seriously disrupt internal order and form.
3. Generally an increase of order over time (owing to an increase in complexity and differentiation of parts).
4. More than mere intake of matter and energy—extensive transactional commerce with the environment.

As this conceptual model depicts open system operations, steps 12 through 20 (Fig. 2) will be explained in detail.

FEEDBACK LOOP OPERATION

The purpose of the feedback loop segment in systems operation is to provide an outlet for energy and/or materials produced by internal systems function and to offer a channel for input of energy and/or materials for further systems use. The feedback loop begins as a systems boundary maintenance function. The exchange of substances occurs across the system boundary. Steps 12 through 20 (Fig. 2) show in detail all that occurs during the momentary exchange in a minute segment of time and space.

The environment outside any system has ecologic variables of influence which impinge on open system operation. These variables are interrelated and could be classified in a variety of ways. The four most general classifications might be as follows:

1. *Mobility Characteristics*—spatial-geographic characteristics, temperature,

availability of essential resources (water, air, nourishment, or chemical and/or basic elements).
2. *Characteristics of the Coexisting Population*—number, type, size.
3. *Physical Characteristics*—presence or absence of damaging elements.
4. *Potency Characteristics*—intensity of any of the above factors.

As this model will be applied to social systems in this text, a more specific classification of ecologic variables would be helpful. Such a list could include as variables of influence spatial, sociologic, cultural, physiologic, psychologic, economic, and political factors.

This environmental input provides influential data from outside the system which through a *transaction* (Fig. 2, step 13) begins to organize the new data into acceptable and/or receivable terms. These data of influence are converted into the individual systems language, or are *encoded*.

The operations found in a feedback loop do more than simply code incoming data. They also control the rate and/or flow of data into the system. *Resonancy, reciprocy, synchrony,* and *helicy* (Fig. 2, step 14) are descriptive terms for the four major types of stimuli rate and/or flow controls. *Resonancy* describes a wavelike stimulus. It is characterized by the effect produced in physics experiments when a natural vibration frequency of a body is greatly amplified by a reinforcing vibration, at the same or nearly the same frequency, from another body. An example of resonancy in an electrical system is when the current or voltage within a circuit is in the same phase as the applied current or voltage. Circadian rhythms found in biologic systems also illustrate resonancy, as does organization of content to be learned by students in repetitive wavelike sequences with play or rest periods in between. *Reciprocy* is a description of mutual interaction between system and environment. The reversible reaction in ester formation

$$C_2H_5OH + CH_3 \cdot COOH \longleftrightarrow CH_3COO \cdot C_2H_5 + H_2O$$

is an example of reciprocy in physical chemistry. Open exchange of information between student, student and teacher in a seminar teaching-learning experience can also be called reciprocy. *Synchrony* describes stimuli presented to and accepted by a system at a specified point in time-space. Discharge of atmospheric electricity from one cloud to another or between a cloud and earth (lightning) is an example of synchrony. Union of a specific sperm and egg bringing about fertilization is synchrony. In a teaching-learning situation, synchrony is when the teacher presents the learner with needed content just when the learner is ready to receive it. *Helicy* describes the cumulative effect of stimuli inputs occurring between system and environment along a spiraling longitudinal axis bound in time-space.[10] Crystal formation and fetal development could be physical system and biologic system examples of helicy. The education process, which extends in this country over a minimum period of sixteen years, is a social system example of this stimulus pattern.

The system components react to these new data (now encoded and controlled in rate and/or flow) through *interdependent interaction* (Fig. 2, step 15). In other words, components which have already been through an initial role definition process that has normalized and provided system coherence and purpose now face additional onslaughts of stimuli to which they must react. Because the role definition process has been jelled at a high level of systems function, components naturally react initially to stimuli in their preprogrammed way. Some chances for role changes can occur if stimuli are sufficiently strong or are well enough directed to overcome role stability. The system may be open to or accept stimuli offered in any single or combination of stimulus forms. The total process of acceptance or non-acceptance, offered or not offered, is called *patterning* (Fig. 2, step 16). As a result of this stimulus patterning, components may change in any or all of the following three major ways: *organization,* the hierarchy of roles; *calibration,* a change in number or amount; and *equilibrium,* a disturbance in the normalizing segment of system function (Fig. 2, step 17). Most often, when change occurs as a result of environmental stimuli, the system changes, at least subtly, in all these three areas.

After such changes either occur or do not occur, the system takes time to *differentiate* (Fig. 2, step 18) that which has now become part of actual system organization from that which is residual from the ecologic influences. *Negentropy* (Fig. 2, step 19) occurs. This is the mechanism by which a system maintains itself in homeostasis at a relatively high degree of orderliness or at a fairly low degree of entropy. Establishment of negentropy comes from the ability to draw orderliness from the environment. Negentropy marks the outer edge of the *system boundary* (Fig. 2, step 20). It is the most steady state of systems function. It is a dynamic equilibrium. Those things not needed for internal systems function are released as output. Products of the systems operation are also present as output. The product is hopefully consistent with the established purpose (Fig. 2, step 10). The remaining energy serves as stimulus on the energy pool, causing the cycle to continue back to step 1 (Fig. 2).

The model presented in this chapter has been drawn in a circular shape so as to enable the reader to view its parts and processes separate from one another. Actually, the external systems segment (feedback loop) and internal system segment intertwine and should appear as in Figure 6. This helical formation depicts energy flow through the system as a continuous, self-perpetuating, self-actualizing force.

Flowing around and through the system enabling it to function internally (intrasystem) and outwardly (intersystem) are the communication networks. Figure 6 depicts intrasystem communication or an intracommunication network as straight line arrows. Intersystem communication or an inter-communication network is represented by the twisting broken line. Both communication networks are necessary for open system functioning. Closed systems have minimal or no intercommunication network. Another way of visualizing a system and its parts is depicted in Figure 7. This figure assists in demonstrating the realities of the processes occurring simultaneously.

The forms these communication networks take depends on the type of system involved. Molecular systems provide for intercommunication networks through interaction of fields of force potentials. Organisms provide parts communication by physiochemical means. Intracommunication of social system components can take many forms from gestured signals and the spoken word to written, verbal, and mathematical symbols. It is necessary that any data coming into the system through the intercommunication network be encoded to the intracommunication networks "language." Outside communication with an atom must be presented in force field potentials; with an organism, in physiochemical terms; and with social systems, in the vernacular of the group.

One of the unique aspects of human communication is that it may also exist in force field potential and physiochemical dimensions. Research in ESP (extrasensory perception) and other "mind over matter" experiences may in time define human communication capabilities in broader terms than the now-accepted stimulus-response feedback arch (Fig. 8).

Communication specialists, philosophers, economists, and psychologists have already questioned the validity of this model (Fig. 8) for human systems communication. The views of such authors as Lee Thayer,[14] Ervin Laszlo,[6,7,8] Kenneth Boulding,[4] and F. Kenneth Berrien[3] relating to social system communication networks will be discussed in later chapters of this book.

That input needs to be encoded into the internal system's language has been discussed. The reverse is also true. What comes out of a system in terms of output (either waste or a product) must be decoded in order to be accepted by linking or interfacing feedback loops. What comes out of a system is not what went on within the system but rather is a result of the intrasystem operations.

The feedback loop and internal system functions operate together in a mutual-causal relationship. Changes in one segment will affect the other.

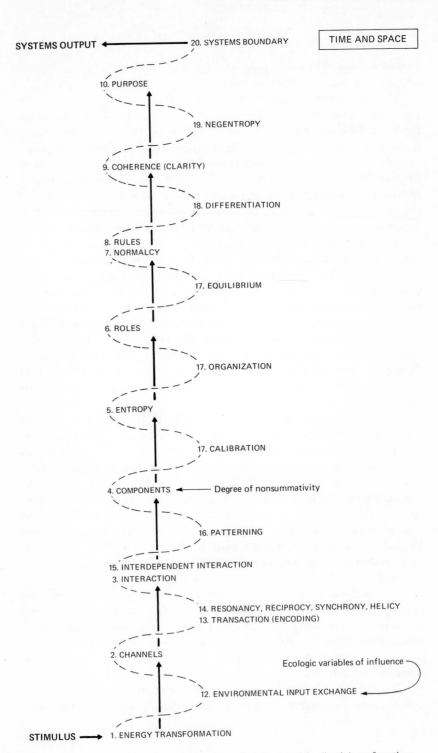

SYSTEMS OUTPUT ◄──────── 20. SYSTEMS BOUNDARY   TIME AND SPACE

10. PURPOSE

19. NEGENTROPY

9. COHERENCE (CLARITY)

18. DIFFERENTIATION

8. RULES
7. NORMALCY

17. EQUILIBRIUM

6. ROLES

17. ORGANIZATION

5. ENTROPY

17. CALIBRATION

4. COMPONENTS ◄────── Degree of nonsummativity

16. PATTERNING

15. INTERDEPENDENT INTERACTION
3. INTERACTION

14. RESONANCY, RECIPROCY, SYNCHRONY, HELICY
13. TRANSACTION (ENCODING)

2. CHANNELS

Ecologic variables of influence ⌐

12. ENVIRONMENTAL INPUT EXCHANGE ◄──

STIMULUS ──► 1. ENERGY TRANSFORMATION

FIG 6: Relationship between internal systems functions and feedback loop functions.

20

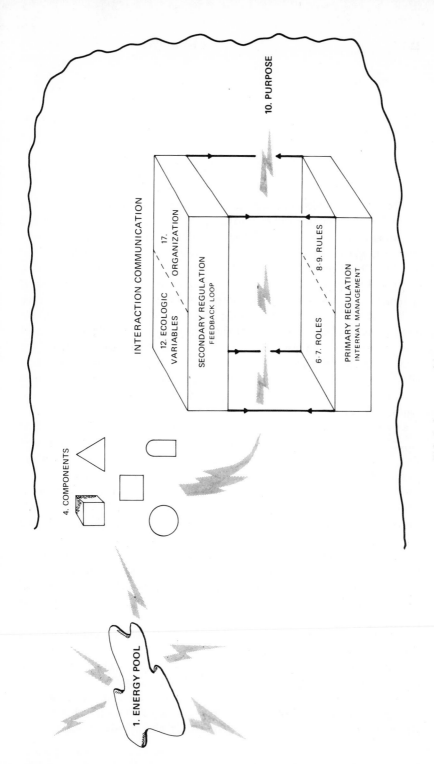

FIG. 7: System model: processes.

21

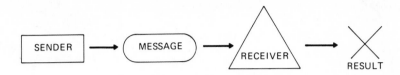

FIG. 8: Communication.

Feedback loops which operate mainly to maintain the status quo of the internal system function will seek to accept only negative feedback which will correct any deviations found in internal system operations. Such a feedback loop arrangement is called a *deviation-counteracting mutual-causal relationship*. The thermostat is a classic example of such a relationship. Feedback loops which operate mainly to increase upon deviations noted in internal system functions will seek from the environment mostly positive feedback data. This type of feedback loop operates in a *deviation-amplifying mutual-causal relationship* with internal system functions. Reproduction of mutant species is an example. The science of *cybernetics* studies these feedback loops by placing value weights on each piece of data selected or rejected by a feedback mechanism. By mathematical calculations, predictions can be made whether the changes made in internal systems function will maintain the current stability or reinforce and add to any divergences.

Cybernetic analysis is a new research tool for the social scientist. It is difficult to apply in complicated systems. The model presented in this text shows the complexities of systems organization. The model as initially presented is a very simple system structure. It is operational at the microsystem level because of the straightforward character of its component element. If each of the components discussed in the model were not a single entity but rather a system in its own right, the model would look like Figure 9.

The components of this mezzolevel system could be called subsystems. Each subsystem has in operation its own feedback loop mechanisms exposed to ecologic variables and its own patterning of stimuli data. The total system, or mezzosystem, also is affected by ecologic variables and stimulus patterning. At the highest and most complex level of systems function, macrolevel, the components may be subsystems whose components are systems. The resulting model appears in Figure 10.

Figure 10, the macrosystem model, shows the complexity of relationships between suprasystems, systems, and subsystems. F. Kenneth Berrien[3] describes what factors influence component systems to develop strongest ties with any of three choices, the larger supra (macro) system, system components on the same level, or smaller sub (micro) systems.

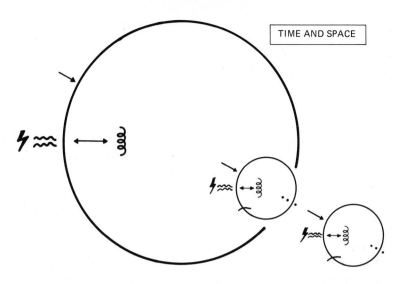

FIG. 9: Mezzosystem model.

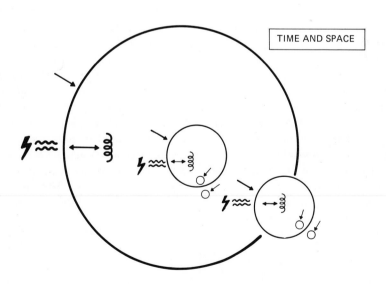

FIG. 10: Macrosystem model.

... sub and suprasystems become interdependent, each supplying appropriate signal and maintenance inputs for the other. In the early stages of such a symbolic relation between sub and supra systems, it is the older subsystem [Berrien sees suprasystems evolving out of growth and increasingly higher level organization of subsystems] that retains the power to determine whether the suprasystem survives and has a greater resistance to destruction. Later, for two reasons, the suprasystem gains the upper hand. First, the growth of the suprasystem is accomplished by capturing within it additional subsystems that become increasingly specialized, and each is less able to survive independently. Second, the suprasystem, taken as a whole, thereby becomes a major source of maintenance input for its components. [3]

This means simply that for a time components of the original system have a greater attraction for one another than for components of subsystems or suprasystems which come into the relationship later through growth and organization. But in the long run, the highest level of system complexity (or the suprasystem) eventually takes over these earlier alliances.

Because of the complexity of systems function, a multiplicity of responses could arise from a single series of stimuli, just as several different stimuli could result in a single response. Thus, the concept of equifinality can be visually appreciated in this model. System assessment by whatever tools used will be a difficult task. Its difficulty and rewards will be explored in Chapter 3 of this text.

Systems and their feedback loops have been discussed in detail, but the space that exists outside systems has not been discussed. This space, or region between the boundaries of systems, is a medium that can transport information, energy, etc., from the output boundary of one system to the input boundary of another system. This medium is called the *interface*. The way the interface can be identified as different from a system is that no interaction between energy, elements, or information occurs while in transit across the space. The interface is a region that accepts entropy. If systems wish to reduce the entropy rate that occurs in the dead space or interface, they will establish links with other systems that frequently interact together. Linkages save time, help direct energy, and facilitate communication, etc., between interacting systems. They also are one of the channels for development of higher level systems (mezzo- or macro-) which can form as microsystems grow more complex.

REFERENCES

1. Allport G: The open system in personality theory. J Abnorm Soc Psychol 61:303-11, 1960, p 303

2. _____: Pattern and Growth in Personality. New York, Holt, Rinehart, Winston, 1961, p 231
3. Berrien K: General and Social Systems. New Brunswick, N.J., Rutgers University Press, 1968, p 92
4. Boulding KE: General systems theory: the skeleton of science. Management Science 2:197-208, 1956
5. Doyle A: The red-headed league. In The Adventures and Memoirs of Sherlock Holmes. New York, Random House (Modern Library), 1955
6. Laszlo E: Introduction to Systems Philosophy. New York, Braziller, 1972
7. _____: The Systems View of the World. New York, Braziller, 1972, pp 68, 75
8. _____(ed): The Relevance of General Systems Theory. New York, Braziller, 1972, p 11
9. Rapoport A: Foreword. In Buckley W (ed): Modern Systems Research for the Behavioral Scientist. Chicago, Aldine, 1968
10. Rogers M: An Introduction to the Theoretical Basis of Nursing. Philadelphia, Davis, 1971
11. Rogers TA: Elementary Human Physiology. New York, Wiley, 1961
12. Schrodinger E: What is life? In Buckley W (ed): Modern Systems Research for the Behavioral Scientist. Chicago, Aldine, 1968
13. Sutherland J: A General Systems Philosophy for the Social and Behavioral Sciences. New York, Braziller, 1973
14. Thayer L: Communication systems. In Laszlo E (ed): The Relevance of General Systems Theory. New York, Braziller, 1972
15. von Bertalanffy L: General Systems Theory. New York, Braziller, 1968, p 211

RECOMMENDED READING

Bertrand A: Social Organization: A General Systems and Role Theory Perspective. Philadelphia, Davis, 1972

Buckley W (ed): Modern Systems Research for the Behavioral Scientist. Chicago, Aldine, 1968

Emery FE: Systems Thinking. New York, Penguin, 1972

Kast F, Rosenzweig J: Organization and Management: A Systems Approach. New York, McGraw-Hill, 1970

2

APPLICATION OF SYSTEMS THEORY

Man is in process, as is the whole of life. He may survive or
he may not, but so long as he survives he will be part of the
changing, onrushing future. He, too, will be subject to
alteration. In fact, he may now be approaching the point of
consciously inducing his own modification.

LOREN EISLEY, 1973
An Illustrated World of Thoreau[7]

Chapter 1 has introduced the reader to general systems theory, in word
and model. Theories and models demonstrate their usefulness through testing
and application. We believe general systems theory is logical and practicable.
We propose to demonstrate this practicality by applying the theory to three
systems: community, family, and health care. This is done not just as an
intellectual exercise, but to increase our understanding of any system. It is
hoped that this understanding will lead to more appropriate and effective
methods of implementing change.

The three systems to be analyzed are interrelated. Family and health care
are both functional subsystems of community. Community has been defined
in a variety of ways ranging from the simple to complex. All definitions
assume interrelationships as basic to the concept of community. To put the

27

human community into perspective, we must examine the past to understand the present. Thus, we will look at culture (why we are), society (where we work), primary groups (purposive associations) such as the family, and health care as separate but interrelated communities.

THE SYSTEMS

Culture-Society

Cultural systems formed when man developed the ability to walk upright, thus freeing his arms and hands for tool using; the brain capacity then enlarged to accommodate use of precise symbolic communication, ie, language. Biologic variables were not the only influences on cultural systems. Man is an animal of place (locality) as well as a biologic organism. He is essentially a surface land animal, bound to geographic land masses. For this reason, the biologic potentialities of the species are the blocks with which cultures are built, and the facts of the geographic space occupied impose a limit on the form culture can take.

Culture is the fabric of meaning in terms of which human beings interpret their experience and guide their actions. All men undergo the same bittersweet life experiences of birth, omnipotence, helplessness, aloneness, togetherness, wellness, youth, illness, old age, death. Cultural systems then have as their purpose the task of organizing and establishing meanings surrounding these life experiences.

A model of the antropologic-cultural system is presented in Figure 1. The components of this macrolevel system (anthropologic-cultural system) are the various subcultural or societal groups found within the influence of this larger system. It is to those social systems that the tasks connected to cultural meaning are delegated. It is the particular social system that decides how the meanings will be practiced. This system also determines the hierarchies of meanings and tasks. Social systems decide what the "dirty" work is and who will do it. They also establish what the most important job is and who will be allowed to do it. As a guide to decision making, each system develops sets of *values*. These values are the normative standards that influence human beings in how they fulfill their role and how they choose between alternative actions. In the context of the larger system, these individual values are translated and transmitted as group ideologies. As the system levels are interrelated, so are the value systems. It is not surprising to see value conflict when man, in enacting his role in one system, must make a decision that conflicts with a value learned or imprinted from another system. These social systems are subject to influence by a large number of variables. They respond

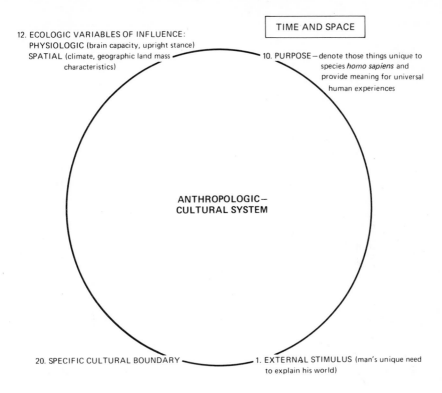

FIG. 1: Anthropologic-cultural system.

to the same physiologic and spatial variables as the larger cultural system, but now that meaning has been given to life experience, variables of economics, politics, psychology, and culture have impact on the societal systems. With inclusion of societal systems as components of anthropologic-cultural systems, the model now looks like Figure 2.

Man is a social animal. Early in man's evolutionary development, he found he was unable to meet his needs alone. Therefore, since the family is the smallest communal unit, it is not surprising that the earliest organizational forms were based on family relationships. Peter Farb[8] has examined Indian groups at their level of social organization. He has developed a taxonomic classification that describes the familial relationships in those tribes organized around bands and the tribe. The tribe is larger than a band, has a larger number of groups, and some specialized functions. The larger groups, bands, and tribes, rather than having political organizations, were organized around secret associations and secret rites. The kinship relationships were important and all members could trace their lineage. Roles were defined as well as rules

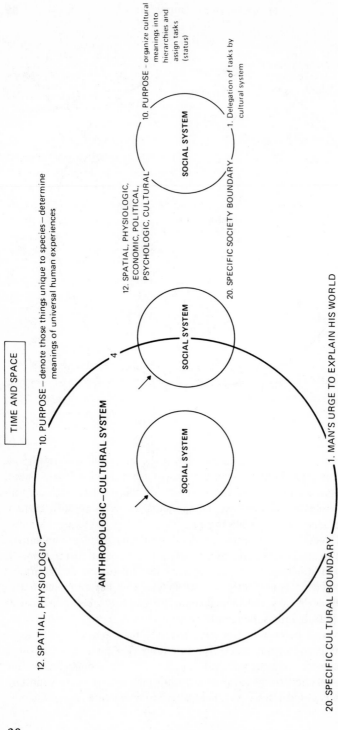

FIG. 2: Relationship of societal systems to anthropologic-cultural system.

TIME AND SPACE

12. SPATIAL, PHYSIOLOGIC

10. PURPOSE—denote those things unique to species—determine meanings of universal human experiences

ANTHROPOLOGIC—CULTURAL SYSTEM

12. SPATIAL, PHYSIOLOGIC, ECONOMIC, POLITICAL, PSYCHOLOGIC, CULTURAL

20. SPECIFIC SOCIETY BOUNDARY

SOCIAL SYSTEM

SOCIAL SYSTEM

SOCIAL SYSTEM

10. PURPOSE - organize cultural meanings into hierarchies and assign tasks (status)

1. Delegation of tasks by cultural system

1. MAN'S URGE TO EXPLAIN HIS WORLD

20. SPECIFIC CULTURAL BOUNDARY

of behavior. Initially, the relationships were formed for biologic survival such as food gathering and procreation.

As areas became more densely populated, the organizations and the relationship became more complex. Europe saw the rise of the feudal system and the Indian counterpart was the chiefdom. Roles became more refined, and in this phase of development came the rise of rank and status. Occupational specialization and relationship to the chief were both important measures of rank and status. This can be seen today, as some occupations are afforded higher status than others. The cheifdoms led to the formation of the state. The essential difference between the two structures was the use of force. In the state, it was the state's exclusive right to administer force. So, in addition to social classes, the state saw the rise of political classes. We have used the Indian society as an example, but the counterparts exist in other cultures; a chronology of cultures will reveal societies that are not related and are distant from each other arriving at the same way of doing things. Anthropologists call this *convergence*; system theorists call the phenomenon *equifinality*. When two cultures come in conflict, there are two methods of resolution: (1) total wipeout, or (2) acculturation with assimilation, so that a new culture develops from a synthesis of the two in conflict.

As man sought to bring order into his communal existence, the organization of the community became more complex. On examination, we see the progression from the headman, chief, and tribal councils, where roles were primarily advisory, to the ruler types as king, to mayor and town councils which have enforcement powers as well as advisory functions.

The roles and rules enacted in the community setting depend on the purpose of the group. Ecologic variables vary in impact dependent on the complexity of the community system.

The earliest community systems (using place of habitat) were those organized around kinship. Even today there are communities that can be defined as extended-family settlements. Some communities were formed to fulfill a single purpose (production of goods), for example, the gold- and silver-mining towns of Colorado and Nevada. Even though the communities did fulfill other of the designated community functions, the economic reasons were paramount. Now these communities exist as ghost towns. Energy was expended for one purpose. When the resources were exhausted, entropy became operative and the towns died. The purpose no longer existed, the people moved on, leaving only reminders of former glory.

Ecologic variables were operative from the beginning. The impact of ecologic variables give a community a unique identity—a flavor all its own. Spatial variables such as geography were often the primary reason for the placement of settlements. The hilltop provided protection, or the river or lake had good drinking water. Economic variables were related to spatial factors. The river or lake was good for fishing, the woods for hunting. Rivers were good for trade routes. Culture, as expressed in religious beliefs, has influenced the establishment of communities, for example, the Mormons in Salt Lake City and the Amish in Indiana. Language, such as French, has influenced the settling of sections of Louisiana.

Culture delegated six tasks to society. These tasks are (1) to meet the biologic needs of members, (2) reproduction, (3) socialization, (4) production and distribution of goods, (5) maintenance of order, and (6) motivation for survival. The more complex the community the greater the number of subsystems (primary groups, associations, institutions) organized to deal with specific functions. These primary groups, because of their specific functions for a defined population, place boundaries on a community that can be identified. These boundaries can be delineated by a knowledge of the purpose or function to be filled and the population to be served. Primary groups provide the individual with a sense of unity through membership. Thus, we hear *our* city, *our* department, *our* country. Early man was limited in his ability to form relationships or to feel this sense of unity with many primary groups other than family and neighborhood. Lack of methods of communication and transportation were hindrances and barriers to expanding relationships. Another example of an old "primary group" was the Guilds. These groups, based on occupational specialization, were forerunners of professional societies and unions. The early universities were known as "communities of scholars."

Some of these primary groups have evolved over the years to where they have almost achieved subculture status. They have their own language, their own values, and rules delineating preferred modes of behavior. An individual belongs to one anthropologic-cultural society, but he may belong to as many primary groups as he wishes.

Family

One of the most complex tasks assigned to societal systems is the socialization of its members. In highly organized complex societies, specialization of subsystems can help to meet many of the socialization needs of individuals. Less organized societies have to depend more heavily on other methods of individual socialization. To date, all known societies have found one method of individual socialization to be the most successful. This method is the family. So to the family is delegated tasks of socialization. The family system is a microlevel system component of a society which can be either mezzo- or macrolevel, depending on its complexity. This societal system is in turn a component of the macrolevel cultural system (Fig. 3.)

The family system is affected by all seven variables of influence and is responsible for the transmission of culturally determined universal life

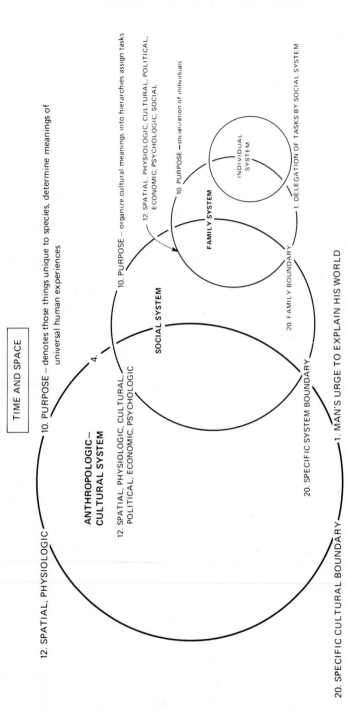

FIG. 3. Relationship of family, social, and anthropologic-cultural system.

TIME AND SPACE

12. SPATIAL, PHYSIOLOGIC

10. PURPOSE – denotes those things unique to species, determine meanings of universal human experiences

ANTHROPOLOGIC – CULTURAL SYSTEM

12. SPATIAL, PHYSIOLOGIC, CULTURAL, POLITICAL, ECONOMIC, PSYCHOLOGIC

4.

10. PURPOSE – organize cultural meanings into hierarchies assign tasks

SOCIAL SYSTEM

12. SPATIAL, PHYSIOLOGIC, CULTURAL, POLITICAL, ECONOMIC, PSYCHOLOGIC, SOCIAL

10. PURPOSE – socialization of individuals

FAMILY SYSTEM

INDIVIDUAL SYSTEM

1. DELEGATION OF TASKS BY SOCIAL SYSTEM

20. FAMILY BOUNDARY

20. SPECIFIC SYSTEM BOUNDARY

1. MAN'S URGE TO EXPLAIN HIS WORLD

20. SPECIFIC CULTURAL BOUNDARY

experience meanings and socially determined hierarchies of meanings (values) and assigned tasks. It is also responsible for specific socialization tasks of (1) provision for sexual experience within limits of incest taboos; (2) provision for reproduction of the species; (3) provision for shelter-protection of offspring during a prolonged dependency period; and (4) provision for the sharing of the work load and produce.

The individual is then a microlevel subsystem of the family system. Because this model has been traced from the largest system level to the smallest; it should not be inferred that the smallest unit has little or no influence on the largest. Such an assumption would be inconsistent with general systems theory conceptualization of equifinality. It is, after all, individuals that comprise the most basic unit of culture, social, and family systems. If the roles assigned to an individual are not reciprocated by him, then some change in role definitions all the way up the line are possible. Where are the individual and family in the cultural-social systems of the twentieth century? What are the major influences on these microsystems and how in turn are they influencing mezzo- and macrosystems surrounding them?

That families function as systems has been well documented by social psychologists working from a social interactionist perspective. Social psychologists and other behavioral science professionals have done an excellent job describing families in terms of internal system functions as well as family feedback loop functions as they interact or interface with other larger systems. Virginia Satir[13] and John Spiegel[14] are two such resources for more family system content.

Because of the unique position the family holds as the major socializing unit of any social/cultural systems, it can have significant impact on the larger systems. The family system can provide shelter and stability for a group of individuals who if left alone in the world would be vulnerable to damaging buffetings. Within the protection of a family system, individuals can heal their hurts, grow strong, and make more efficient use of their available energy. Such efficient use of energy frees the excess to be used to stabilize the higher-level social/cultural systems. Family systems can make society better; however, in the accelerated pace of living experienced today, it is becoming more difficult for families to exert their positive influence.

Toffler[15] describes vividly the impact on individuals of rapidly changing social/cultural systems. He too recognizes that the family is hard pressed to accommodate these changes for the benefit of its members. And he goes on to suggest that perhaps other forms of socialization will be used once the family system breaks down under the stress.

To some extent, Toffler's predictions have already materialized. Nationally, there are many who depend on a larger societal system for the financial resources required to provide shelter, sustenance, and clothing for themselves and their families. In some cases, even though the money is provided, the family is still unable to provide these necessary protective-nurturing functions. It would be easy to explain this inability to provide the basic necessities as the "fault" of those receiving the financial aid; however, the larger system within which they live contributes significantly to this failure. William Ryan[12] explores some of these issues more completely. The goal of professionals working with such families is to improve their functioning level so that they can become more independent from specialized social system services. This task will not be accomplished through isolated treatment of families. Professionals must also be involved in changing the larger systems to accommodate more independent subsystems.

On the other end of the continuum are families that have withdrawn from specialized social systems. These families are seeking to return to a simpler, slower life style and have geographically removed themselves from specialized social systems. Many have either in multifamily groups or single-family units "returned to the land" and are attempting to become self-sufficient in terms of providing their own food and shelter requirements. Some are going further in their independence by educating their own children outside of society's provisions for compulsory education. Although such an arrangement may work well for the adults in these families, there is some concern about children who are socialized to a social/cultural system that "hears a different drummer." There is opportunity for these subsystems to have a large impact on stabilizing mezzo- and macrosystems if they make provision for interaction with these larger systems. The legal system may provide the major means for these family systems to communicate with society/welfare systems.

By far, the majority of family systems fall between these two poles of dependence-independence patterns. With the increased economic pressures of the 1970s, more of these in-the-middle families may be moving to one or the other of the poles. Such movement in itself will have impact on society/culture systems, as the support mechanisms (manpower and money) will be underprovided and overextended. Prevention of such a potentially damaging society/culture system change is another task of the existing social systems.

Berrien[4] describes these interdependence patterns in terms of output. He terms these two outputs FA (Formal Achievement) and GNS (Group Need Satisfaction). Formal achievement is the extent to which subsystems achieve the tasks they are expected to perform. In terms of the family system, FA is evaluated by how well the family meets its socialization tasks. Group need satisfaction has to do with how good the subsystem feels as a result of their interactions with mezzo/microsystems. Families occupying either of the positions discussed earlier have likely experienced a low group need satisfaction. In one case, families have met the formal achievement expectations while families in the other case have also a low FA output.

Development of positive interdependence patterns among the socializing units (family systems) and social/cultural systems is a task of all these system levels. This task can be accomplished through strengthening intercommunication and intracommunication networks. Output in social systems is most often relayed through communication networks. The quality of the output is very dependent on the quality of the communication network responsible. Those concerned with improving either or both families' formal achievement outputs and group need satisfaction outputs work to improve intercommunication and intracommunication networks. The remainder of this chapter deals with family intracommunication networks; the remainder of the book will deal with ways to improve intra-intercommunication networks of one of society system's specialized services—the health care system.

Families follow the systems model in being open or closed. Most of the data about a family's openness or closedness come from study of their intercommunication and intracommunication patterns. As the parents are the builders of a family unit as well as the key blocks in its foundation, their communication patterns are most significant to look at initially. Children learn their basic repertory of communication behaviors from what is presented to them within their family units. Intracommunication networks serve several important functions in a parental dyad. They enable the couple to gather information needed for decision making and make it possible for them to take part in the decision-making process together. Expression of emotions occurs through the communication network. This is a particularly important function for socialization of the children. Children learn where, when, and how to express emotions in a socially acceptable manner. Communication also provides a way for all to strengthen ties with one another. Sharing of necessary information about life in general (not family-related) occurs through communication and provides the means to incorporate input from outside the family unit.

Besides attending to how well couples make functional use of the intracommunication network, it is important to observe actual interactions, because human communication is more complex than message sender-message—message receiver. Even if this model was accepted as being a sufficient explanation, an observer would have immediate difficulty determining exactly what was meant by the words selected to be in the message. Words have different meanings depending on who says them . . . and where . . . and how. The context of the communication, the inflection given the words, and other nonverbal clues are as important as the actual words.

To evaluate the effectiveness of a family's communication network, observe the information-gathering process. Are the data collected clear, specific, and timely? Is the information collected complete in both its emotional and rational components? Is discussion of the collected information also clear, specific, timely, and complete in both emotional and rational tone?

Clarity in communication comes when the verbal and the nonverbal messages match. When a verbal message conflicts with a nonverbal message, the receiver is in a quandary wondering which message to respond to. Most often, in such a situation, the receiver will act *initially* according to the nonverbal message.

Specificity in communication has to do with two verbal messages which conflict. Global generalities in a discussion involving very specific issues will cause confusion and indecision. Giving a compliment and a criticism in the same breath will cause the receiver to hear only the criticism.

Timing of messages is important, as individuals operate on varying tolerances for the reception of outside stimuli (Fig. 2, steps 13 to 15 of the Model in Chapter 1). There are morning people, doves, night people, owls, who have their own time patterns of interacting with others. Are these individual differences allowed and deferred to by the message sender? Also look at the time zone from which the communication is directed. Is the discussion involved with a "now" time focus or is it directed to "past" time or "future" time? Communication dealing with "past" time or "future" time is destructive when "now" time occurrences are never discussed. The same is true if only "now" time topics can be discussed. Elements of all three time zones are necessary for adequate communication, but information focused on should be current and up to date.

Is the message sent complete with emotional tone that matches the content? Or are there filters operating which turn aside anger, resentment, joy, or concern? If filters are in existence, which emotions are let through and which are sifted out?

The previous examples involve the role of the sender and the content/context of the message. A receiver is also involved in the communication process and has some obligations too. Active listening facilitates communication. Provision of a response acknowledging the sent message is also the responsibility of the receiver. A response that says, "I hear you, I'm in touch with you," does much to prevent communication breakdown.

A clear intracommunication network facilitates development of roles that are compatible and reciprocal; rules that are clear and flexible; purposes that are achievable; and boundaries that are both protective and open to incoming data (Fig. 2, steps 6 to 10 of Chapter 1). Such family systems are the ones that provide stability for the members (a high group need satisfaction). These family systems are also able to make society/culture systems better through high formal achievement output.

HEALTH AND HEALTH CARE

Essentially, it is the social/cultural system that places value on characteristics inherent in health or illness. It is also the social/cultural system that establishes roles related to these values. The progression of health care can be traced through identifying the social/cultural systems values about wellness and illness and through describing the characteristics of the associated roles.

Robert Lanon White[17] does just such an in-depth tracing when he describes the most ancient response to serious illness as fear.

> *This first level social response to illness was not based on philosophical conceptions of the worth of the individual, his social position, or any past, present or future contributions to society. Neither was it based on any religious or mystical tenet of spirits, devils, or gods. Rather, it was an individual and collective reflex response to fear of the unknown: avoidance.*

No health-illness roles were assigned by this very early social/cultural system. Those who did become ill were excluded either voluntarily or involuntarily. It was man (the patient) alone against his illness. Although this is not an enviable position, it has inherent dignity.

The same pattern of behavior shows up in later social/cultural systems. White[17] has isolated the essential elements of such a social reaction. Where illness is viewed as a threat, the cause of illness is unknown, and where culturally prescribed remedies do not work or do not exist, fear and avoidance prevail.

Later social/cultural systems began to conceptualize cause and effect relationships regarding health and illness. Animism as a way of explaining or giving reason to everyday occurrences soon became a basis for causes of illness. Those who offended the spirit of the great river, or tree, or sky, or whatever were struck down for their affront. Each man was responsible for keeping himself free of entanglements with the spirit world. If ill, he still was responsible for his own actions and for paying his dues.

Eventually, those individuals who had fewer episodes of illness or who had miraculous recoveries were assumed to have closer ties with the spirits. It was not long before these individuals began to act as intercessors for the less fortunate. This individual or shaman developed the first health-related specialist role. Note the change in role status of the patient, now that he requires someone else to act in his behalf. Over time, as the role of the shaman became more defined and powerful, the role of the patient became more suppliant.

With the advent of Christianity, although the role of the priest-physician is still predominant, there is a change in the role of the patient. The sick now hold an important position in society. In order to do good works, it is necessary to have someone else on whom to work or serve. Who better than the sick and the maimed? As White[17] states, "The Cro-Magnons introduced the humane principle. After 18,000 years the Christians made it a legitimate duty of society."

Although the role of the patient has developed over time to be a subservient one, it has at least for a time some illusions of dignity. The industrial revolution, with its emphasis on doing the job in the fastest way for the greatest production, removed these last shreds. Thus, during our annual physical examination, in the airconditioned office of our doctor, we wear paper coveralls whose exact measurements have been determined by a computer. Also with the industrial revolution came the notion that healthy men work better and are cheaper than sick help. "Health became an economic asset not only to the individual, but for the state as well."[17] It became popular for whole political entities to discuss a state's responsibility for the health and welfare of its peoples.

Health as a broad concept overlaps several of the tasks delegated to society. But what is health? The concept has been pondered and analyzed over the centuries. Health has different meanings to different people. For many, health is merely the absence of disease. Even senior nursing students define health using a variety of terms ranging from "absence of disease" to "optimum functioning of all body processes."[9] Philosophically, is health a state of being—an attitude—a static state, or an evolving, fluctuating process? The World Health Organization struggled with the problem and their definition is accepted, at least in theory, worldwide. That definition, published in 1948, states: "Health is a state of complete physical, emotional, and social well-being, and is not merely the absence of disease and infirmity."* Others like Dubos[5] and Dunn[6] have expressed their belief that the definition is both confining and utopian. Complete well-being is considered unattainable. Dubos believed health could only be meaningful as defined in terms "of a given person functioning in a given physical and social environment."[9] He viewed man in relation to his environment not passively adapting but using his creativity to maximize the process. Halbert Dunn[6] coined the term *high-level wellness*. His definition states:

High-level wellness for the individual is defined as an integrated method of functioning which is oriented toward maximizing the potential of which the individual is capable. It requires that the individual maintain a continuum of balance and purposeful directions within the environment where he is functioning.

This definition depicts man interacting with his environment. But instead of health being perceived as a static or perfect state, health is viewed as a dynamic, changing process. The individual is given the option or mandate to grow, develop, or to improve. Each person can actualize his individual potential within his limitations operating within the framework of Dunn's

*Constitution of the World Health Organization, Chron. of WHO, Vol. 1, 1947. Issued by the Interim Commission. Geneva Switzerland.

definition. That this is possible is demonstrated daily by the thousands of "handicapped" who are functioning above and beyond what might be expected. This relationship (man and environment) discards the medical model of one cause equals one effect. Man-environment relationships denote a variety of variables, apparent both in man and in the environment, resulting in an infinite number of possible interrelationships.

For the individual and family to approach achievement of high-level wellness, the community must provide an environment that is conducive to the achievement of such goals. Society has been delegated health and health care by the larger anthropologic-cultural system. The position of this task in an hierarchy of tasks depends on the value the particular culture places on health. Another factor to be considered is the manner in which society interprets the tasks assigned.

Individual attitudes toward health are a result of family and cultural imprinting, actual encounters with a health care system, and education. Cultural imprinting helps to explain "old wives' tales" and patterns of utilization (hospitals are where you go to die). Some individuals have had unsatisfactory experiences with the health care system (Brand X did not work). The educational system (defined as any organized program designed to change attitudes, values, opinions), through the media, constantly bombard the public in an effort to influence their health-related behavior.

Can health care be categorized as a system or is it so nebulous, so ambiguous that it defies description? It is our belief that health care is a complex, multilevel system.

ANALYSIS OF HEALTH CARE AS A SYSTEM

All systems need energy to function. In describing the health care system, this energy pool (Fig. 4) can be depicted as consisting of millions of individuals, milling about, with amorphous desires and philosophies, undifferentiated, nonverbalized, and nondirected. On the surface, it would appear ameboid, with numerous pseudopodia, but without a nucleus for directed movement. When the lightning flash (stimulus) catalyzes this energy pool, transformation occurs and what was once diffuse and undifferentiated coalesces and crystallizes. Behavior that was once nondirected can now be channeled toward purposeful ends.

What is this lightning flash, stimulus, or catalyst, and where does it come from? Health is a personal matter, each individual defining the concept to his personal satisfaction. Initially, there exists a health need. What are health needs? Do they arise only as a result of some individual crisis that requires treatment, or are there other criteria by which health needs can be

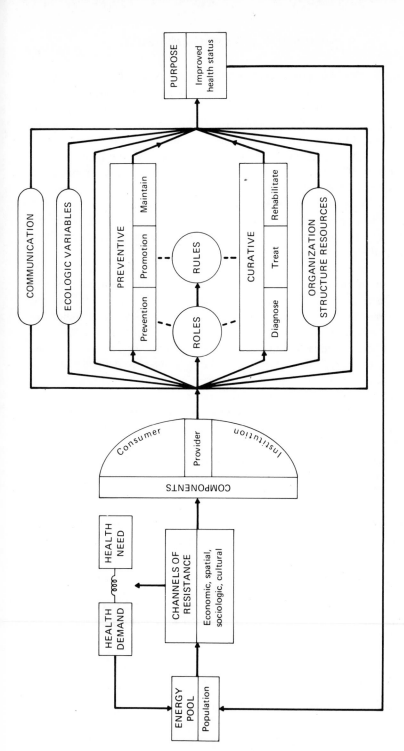

FIG. 4: Health care system model.

41

categorized. The realization that a need exists is not enough; it is the translation of need to demand that provides the stimulus to the energy pool (population). The same factors that impede the translation of need to demand also offer the channels of resistance to further development of the system. Three of the ecologic variables of influence have great impact both as determinants of conversion of need to demand and as continued systems development. These three variables are sociocultural (education level, attitudes, and values toward health), economic (income level, methods available for financing), and spatial (location and type of facilities required.) At first, these needs and demands arise outside the system. Later, as the system becomes self-perpetuating, the stimulus may arise from forces within or outside the system.

The reaction to the stimulus and resultant interaction within the energy pool result in the formation of groups verbalizing their mutual concerns regarding "needs." These groups form the components (Fig. 4) of the system. These groups are analogous to Laszlo's "knots." "Now there is something there—something enduring. . ."[10]

There are essentially three components of the health care system: *consumer, provider,* and *institutions.* The *consumer* is defined (for this discussion) as any person who may at some time require or use the services provided by the system. *Everyone is a consumer.* The *provider* is an individual who has been trained to render those services. An *institution* is a group that functions to make delivery of services possible. The degree of reaction and interaction between these components determines the degree to which the system meets defined needs. The systems term, *entropy* (Fig. 2, step 5, Chap. 1) is consistent with the apathy demonstrated toward needs. When no further organization occurs, needs remain unmet and entropy occurs.

The purpose (Fig. 2, step 10, Chap. 1) of the health care system is to promote and maintain the health of the population. Medical care (here used as synonymous to health care) has been defined as:

> the complete range of personal health services—the promotion of health, the prevention of disease, diagnosis and treatment, and rehabilitation of the patient. These personal health services are produced, financed and delivered through a variety of private and public activities.[11]

Furthermore, since the assumption is that the care provided will be "good care," Myers[11] lists four categories and their constituent elements that will make possible the provision of good health care:

Accessibility	*Quality*
Personal accessibility	*Personal competence*
Comprehensive services	*Personal acceptability*
Qualitative adequacy	*Qualitative adequacy*

Continuity	*Efficiency*
Person-centered care	*Equitable financing*
Central source of care	*Adequate compensation*
Coordinated services	*Efficient administration*

An examination of these four categories demonstrates the interrelationship of internal management operations and the feedback loop mechanism.

Accessibility

Because of the individual definition of health and the priority given to health care, it is difficult to predict when utilization of health services will occur; however, individuals need access to appropriate services when needed. The individual consumer needs a gatekeeper who can be a guide to the appropriate channels by which he can achieve comprehensive care. This gatekeeper should be willing to develop a personal relationship with the client and be able to provide the client with continuity of care. Individually, it is difficult to provide the full range of services required for comprehensive services; however, interinstitutional relationships, networks for liaison relationships, and the maintenance of open communication systems could forward the provision of comprehensive care. Are the services that are provided sufficient to the demand and will they continue to be?

Quality

Quality of care implies that there are standards for education initially and there are mechanisms for continuing education for updating knowledge and skills. For the consumer, it means that the care, in addition to being "good," is acceptable. This infers that the consumer is aware of available options, of professional expectations, and has been given the information to make a wise decision regarding his care. Quality also affects the institutions. Guidelines, standards, and accreditation criteria have been developed for a wide variety of health care institutions to assist in safeguarding the consumer.

Continuity

Myers[11] states:

continuity for the individual involves both a concern for him as a human being in the context of his family and community life, and an orientation toward promoting and maintaining his total health at every opportunity.

This can only be provided if the consumer has a primary provider who can assume responsibility for coordinating the care needed. This person is the consumer's gatekeeper, as he provides access to the system, monitors health status between episodes, and coordinates appropriate referrals. Continuity between institutions infers coordination, planning for efficient utilization without overlap, and joint planning to meet community needs.

Efficiency

Are there mechanisms available that permit the consumer to plan ahead for the payment of medical expenses? Does the provider receive adequate compensation? Are programs planned to make efficient-use of funds and resources available?

As interaction occurs between and within these component groups, the primary regulation processes develop. These primary regulation processes (internal system regulators) are roles and rules (Fig. 2, steps 6 and 7, Chap. 1).

Over time, the system components developed their own characteristic modes of behavior or roles. There are three main role induction mechanisms operative in the health care system. These mechanisms can be classified as the 3-As: ascribed, achieved, and adopted.

The *ascribed* role is the natural one and immutable. As individuals, for example, we are male or female. Thus, in the health care system, we are all members of the portion of the role-pair designated the consumer. In some manner, each day we make use of services made available by the provider, whether it is taking an aspirin at home or taking our children to the pediatrician. Until the late 1950s and early 1960s, this role was largely ignored by providers except as a target for the advertising agent. Consumers obtained services but had little voice in what services were provided, where they were provided, or how they were provided. Providers tended to do things *to* people or *for* people, not *with* people. The consumer was considered functionally illiterate as to his health needs by the provider and the institution.

The *achieved* role is that role for which there exists some prerequisite. In the health care system, that prerequisite is education, either through formal university preparation, as for the professional, or through short-term courses, inservice, or on-the-job training, as with the new paraprofessionals.

The *adopted* role is that role that has been assigned by another. Thus, institutions (here defined as governments, insurance companies, voluntary agencies, hospitals, or clinics) may demonstrate roles that have been adopted rather than achieved. Often this role occurs because of a gap in existing services and either the task is assigned or taken by a group to meet an existing need.

Therefore, each of us has many roles throughout life, dependent upon the life situation. These roles always exist in relationship to other roles. The primary role-triad in the health care system is provider-consumer-institution. Because of the associated rank, status, and power either assigned or assumed by roles, the role triads or dyads, even though the connotation is one of "team," have been interpreted as one role being subservient to the other. Thus, we refer to doctor-nurse, nurse-patient, even husband-wife as if the latter was not equal to the former. This interpretation of subservience in role-pairs even manifests itself within the health care system, where the consumer was usually considered last.

It is not surprising that role conflict occurs when we realize that in addition to being consumers, we in health care services are providers, and members of institutions. Thus, we are often like the lady who possesses many hats trying to decide which one to wear to tea. This conflict is always resolved in some manner; most frequently, it is resolved through denial or preferential selection of one role over one or more roles. Through resolution, the frustration level is reduced to a working minimum.

Another method of resolution is role definition. Thus, new roles were defined and areas of clinical specialization were designed. Some of these were designed for the curative aspects of illness, others for the prevention of disease and promotion of health. Professional jealousies developed and institutions squabbled and competed. The resultant paranoia made communication and meaningful dialogue either difficult or nonexistent between providers in the private or public arenas. Yet each group has contributed much to the development of the complex health care system. Beverlee Myers[11] offers an excellent chronology of the historical development, and this chronology is reprinted in Appendix C of this volume. Collaboration between professionals has usually occurred because of some dramatic event (epidemic or disaster) requiring momentary cooperation. Thus, there exists a "cold war" between the glamorous treatment specialities and the more conventional preventive aspects of health care.

The natural consequences of role definition are the assignment of rank (hierarchy), status, and power. Essentially, this order of dominance defines the relationships between individuals or groups. In the health care system, it appears that those with the most education in specialties dealing with the latest technologic advances have achieved the most in terms of rank, status, and power. Thus, we may perceive the heart surgeon or the coronary care nurse as being the "brightest and best," and those professionals working "with people" as lowest in the pecking order. This type of pecking order is not unique to health care. For example, the electronmicroscopist appears to be of more value than the humble biologist who first described the creature whose giant cells are so sought after by the former. Who assigned the rank order of the health professionals? If human behavior has its counterpart in the behavior of primates, it may be explained by Robert Audrey[2] in the behavior of the "alpha fish." This term as used in *The Social Contract* is the individual who is the dominant figure in a society. It is from this individual that the group takes its direction. Females are seldom "alphas." It has been suggested that this is not the result of an inability to lead but rather of the time required by child rearing. So, in the health care system, physicians as a group, predominantly male and with more education, were given the role of "alphas" by other health providers. It was the physician who allowed the patient to be admitted to the system, stated what his needs were, and who and what would fulfill these needs. In primate societies, the "alpha" accepts with his "alphaness" concurrent responsibilities and obligations. Has the physician fully accepted his role as chief medical provider with its corresponding obligations and responsibilities for provision of health care, continuity of care, and quality of care?

Each person, to function in his role properly, must be aware of the role expectations and the acceptable role behaviors. Thus, an outgrowth of roles are rules. Rules provide role security. The health care system has developed its own unique rules of behavior. The consumer until recently was given the rules governing the passive recipient of services. These rules were unwritten and changeable. The provider had a wide variety of rules that are evident through practice acts. As clinical specialties developed, criteria for admission and qualifying examinations for candidates were developed, as well as criteria for excellence of practice. Institutions developed procedure manuals. As more subgroups evolved, institutions developed job descriptions to delineate rank and function within the staff groups. Other rules are unwritten. Many of these hidden or covert rules, when analyzed, have been found to be based on tradition and are illogical and formulated from information that is currently invalid.

As self-actualization is an innate human need, an open system is self-actualizing through achievement of purpose. Self-actualization of a system occurs through the simultaneous processes of internal system management and the secondary-regulation processes (feedback mechanism). Feedback systems exist for the regulation of actions. The chief elements of the secondary regulation processes are organization (Fig. 2, step 17, of Chap. 1) and the ecologic variables of influence.

Organization provides the system with structure, an operating framework for the achievement of purpose and ultimate self-actualization. It is through organization that decision-making bodies provide for the division of labor and the allocation of resources to fulfill system purpose. If, philosophically, the goal or purpose of the health care system is provision of services whereby individuals and communities can reach their maximum potential for wellness, how then has the system organized to fulfill the purpose?

In the beginning, when societies were simple, the priest or shaman was responsible for organizing the delivery of health services and setting or interpreting health rules and regulations. As communities and societies became more complex, diversity of services was required, and more structure was necessary. The overall purpose was analyzed and the system components became more specialized in their interpretation of their functional roles to meet the goals. Thus, we have the curative sector (diagnosis, treatment, rehabilitation) and the preventive sector (health promotion, prevention, maintenance). Each sector has organized in its own unique fashion to add complexity to an already complex situation.

The curative sector, based on the medical model, can be seen as a series of independent medical practitioners and their subordinate professionals and nonprofessionals, with their offices or hospitals as their scene of action. The main basis for interrelationships is peer groups and independent professional groups. Evaluation of practice is only accomplished through peer review of hospital records and hospital committees such as tissue or infection committees that monitor the number of operations, days in hospital, postsurgical infections, etc. The consumer, as a patient, is rarely consulted as to the modality of treatment, hospital setting, or how he feels about his course of treatment and care. Certainly, the consumer evaluates his care, but his only recourse, if he feels he is maltreated, is a complaint to the local medical society or to the hospital administrator, or in extreme cases a legal suit of malpractice. The consumer's dissatisfaction results in doctor hopping, with the result that continuity and coordination of his care suffer.

In contrast, the preventive sector can be viewed as a highly structured network that interrelates vertically (local to international) as well as horizontally (interstaff). The exact structural form varies from level to level. Many organizations are presently struggling with reorganizational plans to make their programs more relevant to the consumers in their areas. Each level has designated functions and responsibilities that are common for all units at that level; however, implementation to fulfill these obligations varies from place to place and demonstrates responsiveness to local need and the creativity of the provider.

Concurrently, there exists in the preventive sector groups organized as voluntary associations rather than official ones. These associations have developed in response to specific community needs. They often accomplish a great deal because they are innovative and are not hampered by bureaucratic red tape. The sad thing is the lack of coordination between official and voluntary groups, which results often in an overlap of services; however, even with these additional voluntary associations, there are still gaps in health services.

Even though the consumer is sought out and care provided for specific needs, he may still be confused and frustrated regarding which area has the ultimate responsibility to meet his needs. Contributing to this problem has been the image of public health as a "controller of disease" rather than a primary force for health promotion. The image of the preventive sector has been further blurred by the interpretation that instead of providing services for all the public, the available services are only for the poor. Thus, the consumer, who does not wish to consider himself poor, assumes that the services he wishes or needs are not available to him.

Ecologic Variables of Influence. These are the elements in the environment that influence our behavior. The variables are multiple and interrelated. A change occurring in one variable consistently affects other variables. These variables may be conceptualized as in Figure 5. Even as they interface with the system, so they also interface with each other, forming a network around the system. These variables, in terms of specific aspects, have changed over the centuries. Each generation has colored the ecologic variables with its own

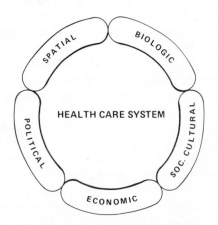

FIG. 5: Ecologic variables and the health care system.

characteristic pattern. Dubos[5] states that "the complex nature of man's response to his environment accounts for many of the difficulties experienced in developing methods for the prevention of disease. . . ." These ecologic variables (spatial, biologic, sociocultural, economic, and political) and their impact on the system will be discussed separately, but they affect the system simultaneously.

Spatial Variables. These are the geophysical factors that affect the system. In reference to the health care system, we are concerned primarily with geographic boundaries. Certainly, other geophysical factors such as climate and season affect health, but it is the geographic boundary that has greatest impact on the health care delivery system. The geographic placement of the system is important as well as the geographic area for which that system is accountable. Health care systems, whether the focus is curative or preventive, are much more complex within urban metropolitan areas. This complexity is a result of several factors: more consumers, an increase in number as well as the type of provider, and the fact that urban living brings about an increase in types of health needs. The newcomer to the city is at once overwhelmed by the number of services available, confused as to where to seek appropriate services, frustrated by the lack of coordination, and completely in awe of his city cousin's seeming ability to get through the jungle and obtain care. Health care in a rural area is much simpler, not necessarily because of geographic limits, but because of fewer and less diversified providers. Therefore, unmet needs exist. A lack of referral resources causes frustration and apathy in both the consumer and the provider.

Biologic Variables. These are the variables such as age, sex, genetic makeup, and disease entities. Biologic variables have great impact on the health care system. Advances in both areas of health care have affected and are affected by these variables. Certainly, within the context of this variable, each generation has had its unique characteristics. Man lives longer because of advances in preventive and curative health care. The causes of death in earlier centuries, the plagues and pestilence, have either been completely eradicated or placed under control. But to take their place have come the modern problems of population growth, problems resultant of chronic diseases, and problems of aging, all of which have challenged our creativity, for the past holds no clue for solutions; after all, in the past there were few who lived past forty. With the changes and the adaptability of human beings, as individuals and groups, to ecologic change, we can only hypothesize what is held in store for the next generation. The gene pool remains the same but the expression of the gene is altered by the environment. We have seen this demonstrated by comparison of present to past children through "normal" height and weight measurements. What will be the impact of our present environment on our progeny? What problems will be presented to the health care system of the future? For the community, a knowledge of the biologic variables can have impact on program priorities. It is reasonable that top priority is given to older age groups in Clearwater, Florida because of the high percentage of retired persons in the population.

Sociocultural Variables. Culture is the sum total of our experiences—the language, values, and attitudes—our "way of life." Aliens who arrive at our shores are said to be acculturated when they leave old habits and take on the American way of life. An effective health care system takes into consideration the attitudes, values (culture), of those it is purported to serve. The success of a comprehensive health program for children in a predominantly Puerto Rican neighborhood resulted in part from the presence of. Spanish-speaking providers. Many other programs have had successes in using the neighborhood worker as part of the "team," as client advocates to assist in explaining the values and attitudes to the staff and in explaining the program to consumers. Attitudes, toward themselves, health, and the health care system can become barriers to communication between consumer and provider and prevent the provision of appropriate services. An attitude on the part of the provider that lacks understanding of other ways of living except that of the white middle class can again prevent proper utilization of services provided. This is not to say that we all have to be psychologists or sociologists in order to function effectively. Rather, an attitude of mutual respect, learning, and humanism can do much to further relationships between consumer and provider. The curative section of the health care system gives token lip service to the sociocultural variable; however, in education and practice, only momentary emphasis is placed on the relationship of these variables to the response of the patient to either his illness or his treatment. The preventive system fares little better. Its members make an attempt, but it is a learning-through-experience that is often difficult for the consumer as well as for the provider. The nutritionist who invested her time and effort to translate the ethnic food preferences of an elderly Mexican woman into a diabetic exchange diet not only demonstrated creativity but a willingness to understand this woman's way of life. The following poem, written by an unwed pregnant teen-ager, demonstrates how the attitudes of others affected a life and the resultant effect on her attitude.

No Better Than Me

I am only as human
As Nature allows,
Governed by virtues
and morals and vows,
Doomed to be judged
By persons I see,

All in God's eyes
  No better than me.
Followed by snickers
  And comments and stares,
I try to pretend that
  I really don't care,
Carrying a child
  That's destined to be
Doomed in their eyes
  No better than me.
My mind has matured
  As my judgment has grown.
I know now I never
  Have once stood alone.
God has opened my eyes
  And now I can see
That those who must judge
  Are no better than me.

Economic Variables. Who pays, for what, and how much? The health care system is expensive. Currently the curative aspect of health care is primarily supported through private payment. Other means of payment are through some form of "third-party" payment such as private insurance, medicare or medicaid. At present, however, "third-party" payment is neither total reimbursement nor all-inclusive for services. Few insurance companies cover the newborn baby until he is two weeks old. A family who has a child born with a birth defect requiring immediate treatment could easily be bankrupt before the baby was even eligible for insurance coverage. Dental services, through private coverage, are included only if the condition is the result of an accident. Medicare and Medicaid are designed to assist specific segments of the population in their medical treatment. Moreover, there are hospitals and physicians who refuse to treat the patient whose only manner for payment is one of these plans.

The preventive aspects (official agencies) are primarily supported through allocation of the tax dollar. Thus, all of us participate in the funding of preventive health organizations. This tax dollar, whether allocated locally or nationally, is distributed on the basis of priorities that are set by the funding agency. Thus, each year, agencies receiving the tax dollar raise their voices in the hope that by speaking loud and long they can be given a higher priority and receive more money. Granted, official agencies have been assigned specific functions and responsibilities, but just how far will the available money stretch? As the dollar buys less at the grocery store, so the dollar as a health resource buys fewer services. As fewer dollars are available, increased emphasis is placed on evaluation of services. Money is spent on essential services that have demonstrated effectiveness, but, at this point, if it has not been possible to measure the impact of a program it has been abandoned. Local health departments are caught in the paradoxical situation of existing to provide services for all, but actually only being able to provide services for the poor. Is health care a "right" for all or only a "privilege" for those who can pay the bill? How much responsibility should the federal government assume? The debate over some form of national health insurance still continues. In 1948, Shirley Basch[3] compared the arguments used against free public education with those being used against national health insurance. The same arguments are used today by the opponents of the latest bills. Odin Anderson,[1] in his historical development of the issue of national health insurance written in 1951, was overly optimistic in his belief that such a health insurance program would be enacted that year. Needless to say, 25 years later such a program is still under debate.

Political Variables. Political, here, is used to denote governing bodies, regardless of party affiliation; however, a historical review will reveal basic party differences in attitudes toward the health care system. Governing bodies certainly have great impact on the system. Most often they exercise control over how the money will be spent. It is to the governing body, whether it be city council, health board, state legislature, etc., that the provider sends his requests. The provider then sits, waits, and worries (and prays) that this body will see fit to grant his requests and he will not only continue to function adequately but will be allowed to further his work. The supplicant hopes that his proposal will be so well written, so well presented, the justifications so clearly identified that the governing body will pass favorably on his request without question. Grantsmanship and proposal-writing has become an "art." Knowledge of what governing bodies "like" in terms of requests, formats, grammar, etc., is stored by "old timers" and passed along to "new" supplicants. Time was when requests covered a multitude of areas. Now in times of tight "monies" the provider has had to assume the subrole of "competitor" in the arena for funds. Around council tables the battles rage in the war of priority for funds. Each proposal is scrutinized for appropriateness, applicability, and justification of need. For too long, the health provider has kept his head in the sand and denied political impact, but no longer, for it is the governing body that controls the purse strings. What do we know about this governing body? What are their attitudes and values relevant to health care? Can they act objectively on matters pertaining to health? Are they, in their business lives, slum landlords or businessmen with plants that are polluting air or streams? Are the health proposals, if passed, going to injure their pocketbooks? If we are part of the problem, we are also part of the solution; we can participate as consumers and providers to assist the governing bodies in their decision making. We do this as providers by giving them all the facts and data available; we do this as consumers and providers by participating in open meetings and hearings.

The key to effective systems functioning either intrasystem (between components) or intersystem (between systems) is interaction or *communication*. There are three essential reasons for communication breakdown in the health care system: lack of common language, educational experience, and the concept of territoriality.

Each professional within the system has developed his own language that ensures his niche in society. This language has helped give him identity but it has also hindered communication between providers as more varieties of professionals have become interested in health care delivery. Each professional brings to any interdisciplinary discussion his own language. Unless time is devoted to clarification of terms, deliberations can end in deadlock and frustration. The consumer often has difficulty understanding much of what he is told, and professionals often find translation difficult. Is it any wonder that treatment regimens are not followed when consumers do not understand what is expected?

As providers have striven to achieve professional status through education and the compilation of a "defined body of knowledge," they have been pulled further apart. There are few opportunities within the educational experience for joint learning. This leads to inappropriate role expectations. Often the role expectations are such that communication is impossible. The formation of specialized language and confined learning experiences lead to the development of professional territoriality. Territoriality has been defined as an innate drive to seek and defend an area of space or action. Within the health care system, individual groups of providers are often so intent on guarding "what is theirs" that the patient/client becomes lost and appropriate care is not provided. Territoriality hinders and inhibits collaborative action.

Because of these three factors, it is difficult for all concerned to verbalize mutual needs and set mutually acceptable priorities. The impact of these variables is not always the same; however, the variables are interrelated in their impact on the system. For example, funding (politico-economic) may be dependent on a specific area (spatial), for a program aimed at a specific target population (sociocultural) to solve a particular problem (biologic).

We have examined some significant aspects of the health care system. Criteria are met to qualify health care as a system. The final diagnosis as to type of system, open or closed, we will leave with the reader. Each community is unique. Therefore, some systems, depending on the interacting variables, may be more open than other systems.

REFERENCES

1. Anderson OW: Compulsory medical care insurance, 1910-1950. From The Annals of the American Academy of Political and Social Science 273:106-113, January 1951. In Schuler E et al (eds): Outside Readings in Sociology. New York, Thomas Y Crowell, 1952
2. Ardrey R: The Social Contract. New York, Dell, 1970
3. Basch S: The pains of a new idea. From Survey Graphic 84:78-79, 1948. In Schuler E et al (eds): Outside Readings in Sociology. New York, Crowell, 1952
4. Berrien FK: General and Social Systems, New Brunswick, NJ, Rutgers University Press, 1968
5. Dubos R: Man Adapting. New Haven, Yale University Press, 1965
6. Dunn H: High-Level Wellness. Arlington, Va., R.W. Beatty, Ltd, 1961
7. Eisley L: Afterword. In Chapnick H (ed): The Illustrated World of Thoreau. New York, Grosset & Dunlap, 1974
8. Farb P: Man's Rise to Civilization as Shown by the Indians of North America from Primeval Times to the Coming of the Industrial State. New York, Dutton, 1968
9. Herban N: Student Perceptions of Community Health. Unpublished data
10. Laszlo E: The Systems View of the World. New York, Braziller, 1972, p 81
11. Myers B: A Guide to Medical Care Administration. Vol. I, Concepts and Principles. New York, American Public Health Association, 1969
12. Ryan W: Blaming the Victim. New York, Vintage (Random House), 1971
13. Satir V: Conjoint Family Therapy. Palo Alto, Science and Behavior Books, 1967
14. Spiegel J: Transactions. In Papajohn J (ed): New York, Science House, 1971
15. Toffler A: Future Shock. New York, Random House, 1970
16. Watzlawick P, Beavin J, Jackson D: Pragmatics of Human Communication. New York, Norton, 1967
17. White R: Right to Health: The Evolution of an Idea. Iowa City, University of Iowa, 1971

RECOMMENDED READING

Anderson R, Carter I: Human Behavior in the Social Environment. Chicago, Aldine, 1974

Biddle B, Thomas E (eds): Role Theory: Concepts and Research. New York, Wiley, 1966

Colt A: Public policy and planning criteria in public health. Am J Public Health 59(a):1678-685, 1969

Cooley C: Primary groups. From Social Organization. New York, Scribner, 1909, pp 23-31. In Schuler E et al (eds): Outside Readings in Sociology. New York, Crowell, 1952

Homans G: Groups and civilization. From The Human Group. New York, Harcourt, 1950. In Schuler E et al (eds): Outside Readings in Sociology. New York, Crowell, 1952

Jacob F: The Logic of Life: A History of Heredity. New York, Pantheon, 1974

Kast F, Rosenzweig J: Organization and Management: A Systems Approach. New York, McGraw-Hill, 1970

Keesing F: Cultural Anthropology: The Science of Custom. New York, Holt, 1966

Kluckhohm C: The concept of culture. From Mirror for Man. New York, McGraw-Hill, 1949. In Schuler E et al (eds): Outside Readings in Sociology. New York, Crowell, 1952

Linton R: The Tree of Culture. New York, Knopf, 1955

Mechanic D: Sociology and public health: perspectives for application. Am J Public Health 62(2):146-51, 1972

———: Ideology, medical technology, and health care organization in modern nations. Am J Public Health 65(3): 241-47, 1975

Reverby S: A perspective on the root causes of illness. Am J Public Health 62(8):1140-42, 1972

Robischon P, Scott D: Role theory and its application in family nursing. Nursing Outlook 17:52-57, 1969

Thomas W: The primary group and the definition of the situation. From Volkart (ed): Social Behavior and Personality. New York, Social Science Research Council, 1957. In Schuler E et al (eds): Outside Readings in Sociology. New York, Crowell, 1952

Wallis W: Geographical environment and culture. From Community: A Sociological Study. New York, MacMillan, 1928. In Schuler E et al (eds): Outside Readings in Sociology. New York, Crowell, 1952

Yura H, Walsh M: The Nursing Process: Assessing, Planning, Implementing, Evaluating, 2nd ed. New York, Appleton, 1973

Section II
COMMUNITY HEALTH DECISION-MAKING

It's knowing what to do with things that counts

Robert Frost

"At Woodward's Gardens"

INTRODUCTION

Coupled with the increased emphasis on community health and community-based or ambulatory services is the demand for quality care. The public, whether consumer or provider, wants its money's worth. The demand for greater accountability makes imperative effective, efficient health care management. With these cries for available, appropriate, and comprehensive services becoming more strident, is there a simple model that can be utilized in an attempt to solve this complex problem? What approaches, available resources, and usable tools will assist us in reaching our goal of high-level wellness for the community? Problem-solving by anticipating the event rather than solving problems by dealing with what has already occurred should be our aim. This anticipatory approach subsumes a dynamic, creative decision-making process.

A variety of authors have analyzed the decision-making process and have designed a variety of models ranging from the three-step model of Simon[5] to the more complex model of Gore.[3] We are proposing a four-step model for decision-making in relationship to an identified system (Fig. 1).

Community health nursing has been utilizing, in the care of individuals and families, what is called the nursing process. This process is a cybernetic method of problem-solving that works with client/patient systems to monitor and regulate information transfer between systems and their environments. It essentially seeks to identify the variables of influence on a given client/patient system and to aid the system in changing its feedback loop processes to accommodate or to block specific variables. It is a dynamic continuing process that leads to patient/client participation in health decision making and when used appropriately includes the element of quality control. The

58

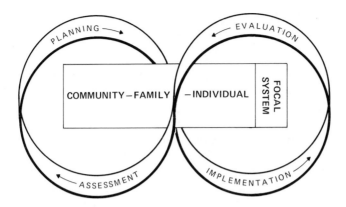

FIG. 1: System decision-making model.

steps in the process (assessment, planning, implementation, evaluation) are not unique to nursing and are appropriate for use by other disciplines as well. The continuous, interwoven phases as depicted in the model demonstrate decision-making as a dynamic response to the focal system. Each step in the process requires a series of choices—some decision to be made. These same four steps which work well with individuals, families, and small groups work just as well when applied to the larger focal unit of community.

The process consists of four phases: *assessment*, containing the elements of data-collection, analysis of data, and statement of conclusions; *planning*, with the elements of ordering priorities, defining objectives, and selecting the appropriate solution; *implementation*, the action; and *evaluation*, measurement of actions.[6]

The focal system has three elements: place, persons, and time. It is essential to define these constants to place plans, actions, and evaluation in the proper perspective. Health is a dynamic variable affecting persons, who live in places, at some designated point along the continuum of time.

1. Place. How are we defining our focal community? What are our boundaries? Is it a six-block area in an inner city, a state, or a larger area that may cross several political boundaries? The focal system to be considered in the remaining chapters will be the community as a mezzolevel system existing in time and space. The National Commission on Community Health Services[7] coined a phrase, "community of solution," as a result of difficulty finding a consensus definition for "community." Their use of "community of solution" makes it possible to transcend political and spatial (geographic) boundaries if necessary to deal with a health issue. They suggest that the boundaries of a community of solution are established within the context of describing, acting upon, and remedying a specific problem. The focal system as pictured in Figure 1 could also be called a "community of solution."

Such communities of solution can exist in either or both of the dimensions of functional communities and structural communities. Functional communities are nongeographic (occupational, religious, special interest, common need, or common resource) aggregates which contain the element of a place of belonging or personal identification for those involved. Structural communities are usually organized by spatial/political boundaries. They are collectives that can exist in an organizational community, such as an inpatient hospital setting; in a primary or folk community, such as a housing/apartment complex, neighborhood, parish, or ghetto; or in a legally established community called a village, town, city, county, state, or nation. Bergwall et al[1] subdivided this term into seven subcommunities. These communities are as follows: (1) *Community of Identifiable Need*: This community is characterized, usually, by geographic boundaries; however, it could be a special population such as migrants. Identified need is defined as a lack that society feels is worthy of attention; (2) *Community of Problem Ecology*: Some conditions when analyzed reveal that defined boundaries are useless for planning. This type of community requires cooperation between a variety of institutions, etc. Such an example might be water pollution of a river that affects all the communities below the level of the source of pollution; (3) *Community of Concern*: The problems affecting the community of need may "spill over" into another area. Therefore, this area becomes part of the community of concern. An example is an outbreak of a communicable disease; (4) *Community of Special Interest*: These communities cross over many geographic barriers and are composed of groups expressing a special interest in a problem. Examples are special organizations such as those formed to combat multiple sclerosis or birth defects; (5) *Community of Viability*: These are the communities that have enough clients to make the activity flexible; (6) *Community of Resources*: This is the community that has the resources, men, money, or facilities to accomplish the task. Needless to say, communities (1) and (6) are not always the same; (7) *Community of Action Capability*: This is the community having the potential or authority for effective follow-through to correct the identified problems. R. Buckminster Fuller[2] treats Earth as a "community of solution" incorporating all the preceding categories. It is easy to see that any "community" classification is neither all-inclusive nor all-exclusive. To implement the nursing process within a community, it helps to be able to describe the initial focal system or "community of solution" in terms of its functional and structural characteristics.

2. **Persons.** Persons are the population of the defined "place." People are the living components of the system which give shape and form to the focal community. It is to people that events happen.

3. Time Time is now! Definition of the time and space influences on the focal system will also help establish the boundaries of the community of solution. Einstein discerned that time is relative to the individual observer. It would be important to understand the focal system's perception of time or its time frame before beginning the community decision-making process. Time frames can be looked at in terms of speed—how fast or how slow do people move in this community? Or in terms of zone—do people in this community ruminate over the past continually (past-time zone); live for the moment only (now-time zone); or live anticipating always the future (future-time zone). The time may be now for those seeking to intervene with a community but it may be yesterday or tomorrow for the people within the community. Space in its broadest sense has to do with the stability of the system's existence. If the "community of solution" or focal system involved is very erratic, it may not actually exist in space long enough to be dealt with. Space and time can never be considered independent of one another. Space is never static. It is not possible to explore it in terms of static-space relationships. Time relationships exist; static-space relationships do not. Thus the "community of solution" either does or does not have stability in space over time. Those that do not cannot be managed. They entropy before new energy input can be helpful to them. Man's perception of time and his knowledge of cyclical change is a tool to be utilized for the future. We do, indeed, build for today and tomorrow on the happenings of the past. We live in the present, reflecting on the past, and imagining the future.

The focal system needs to be considered initially in terms of time orientation, stability in space, and grouping of people. Such consideration will not identify in depth the "community of solution" but will narrow it to a ten-acre field.

In summary, health decision-making in a community consists of four phases: assessment, containing the elements of data-collection, analysis of data, and statement of conclusions; planning, with the elements of setting priorities, defining objectives, and selecting the appropriate solution; implementation, the action; and evaluation, measurement of actions. Operationally, these four steps are continuous and at times occur simultaneously. It is not unreasonable to implement a plan to meet an immediate goal while assessment continues to enable long-range planning. The evaluation process is ongoing to provide continuous feedback for reassessment and reordering of priorities based on new data.

These four phases will be discussed in Chapters 3, 4, and 5, and consideration will be given to a variety of tools and resources. Each phase will be considered as though it occurs as a separate, discrete stage; however, in reality, there is overlapping of stages, as phases occur simultaneously.

REFERENCES

1. Bergwall D, Reeves P, Woodside N: Introduction to Health Planning. Washington, Information Resources Press, 1974
2. Fuller B: Operating Manual for Spaceship Earth. New York, Simon & Schuster, 1970
3. Gore W: Administrative Decision-Making: A Heuristic Model. New York, Wiley, 1964
4. National Commission on Community Health Services: Health is a Community Affair. Cambridge, Harvard University Press, 1967
5. Simon H: Administrative Behavior. New York, The Free Press, 1957
6. Yura H, Walsh M: The Nursing Process. New York, Appleton, 1973

RECOMMENDED READING

Folsom M: Executive Decision-Making. New York, McGraw-Hill, 1962
Kepner C, Tregue B: The Rational Manager. New York, McGraw-Hill, 1965
Lindblom C: The science of muddling through. Pub Adm Rev 19:79-88, 1959

3
ASSESSMENT

Webster defines "assess" as to estimate or determine the significance, importance, or value. Assessment is neither instant nor static, but a continuous evolving process. Assessment is the first step in problem identification and the initial step necessary for sound planning. In theory and practice, assessment is a means by which the assessor, by drawing on the past and the present, is able to predict or plan for the future. Some authorities include assessment as part of the planning phase. We believe that assessment is of enough importance to be considered a separate entity. Assessment is the planning undertaken to define or delineate need, whereas planning, as a separate phase, is planning for action to meet the defined need. Assessment contains three elements: collection of data, analysis and presentation of data, and statement of conclusions (Fig. 1). This chapter will be concerned with the tools used to identify specifically the "community of solution" as it relates to a defined problem.

Within the systems framework, assessment is used to describe the process utilized to determine strengths or weaknesses or to define system tension and stress. Such descriptions of tension and stress come from assessment of the internal process/structural arrangements of the focal system. The systems model (Fig. 2, steps 1 to 10 of Chap. 1) shows that internal system assessment includes characterization of the stimulus, resistance behaviors, component identification and degree of nonsummativity, roles and rules seen, degree of clarity in these roles and rules, and the stated purpose of the system. As we are assessing social systems, most of these data are collected through observation of behaviors. Descriptions of tension and stress come out of observation on how interrelated the components are. Does the right hand know what the left is doing?

63

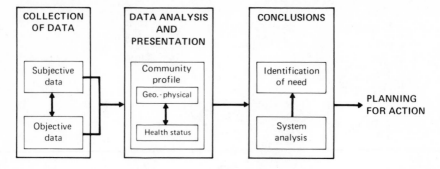

FIG. 1: Assessment model.

　　　Tension and stress can also be identified through observation of role expectations and behaviors within the system. Is the hierarchy of roles that has been established functional and comfortable for the components? Are there components seeking to reject or change role definitions for themselves and/or others? Other ways of determining internal system stresses involve identification of the system's rules. Are the rules clear and can they be presented openly? Can the rules only be learned through breaking them? Last, is the stated purpose of the system consistent with the behaviors observed? Are there any unstated purposes? Does everyone within the system know what the purpose is?

　　　System assessments also need to include feedback loop process/structural arrangements. The feedback loop handles boundary maintenance tasks and its specific processes work to define the system's openness or closedness to new data. Steps 12 to 20 of the systems model (Fig. 2 of Chap. 1) illustrate these operations. Which of the ecologic variables are allowed to influence the system? Identification of the system's definition of deviancy is important to know, as the labeling process is a major factor in determining what variables can or cannot enter a system. A more in-depth discussion of this process occurs later in this chapter. Description of the system's responsiveness to particular stimulus patterns is important too. Does the system respond more openly to data presented in waves (resonancy), such as receipt of letters, phone calls, and personal contact? Consider the political system as an example of one which usually responds most openly to the resonancy stimulus pattern. In order to have impact on our government representative, we are urged to write letters, send telegrams, phone, and use lobbyists. The lobbyists also use reciprocy and helicy stimulus patterns through providing information to the congressman on specific variables upon request and through maintaining availability over time. Some systems respond most readily to synchrony and are open to receive data only at very specific times. Many of our bureaucratic systems have this reputation because of their red tape requirements. It is also possible to observe, through the feedback loop operations, which system point is the most receptive to change. Are components added or subtracted often? Is the organizational structure always being "updated?" Is change the only constant when considering equilibrium? Add these descriptions together and it should be easy to determine if the system boundary is relatively clear or unclear; open or closed.

Systems exist within systems, so for an assessment to be complete it is useful to assess one system above and one below the focal unit. For example, the family system can only be comprehended in the context of the larger societal system and with consideration of the smaller individual system components. Community systems need to be described in terms of their smaller subsystem components as well as their larger suprasystem position.

Finding the internal stresses of system units above and below is relatively uncomplicated. F. Kenneth Berrien[1] describes the process of internal conflict between systems in terms of growth. As subsystems of a system gain in strength and specialization of function, the larger suprasystem takes over alliance ties. During the period of alliance changeover, subsystems experience stresses associated with loss of established interaction patterns. Younger, less specialized subsystems show more affiliation with component systems at the same level and maintain loose associations with the larger suprasystem. The central issue in internal conflict then has to do with the tendency for suprasystems to display excessive dominance over subsystems at the expense of the subsystems involved. Growth is an inherent characteristic of systems, and for systems to remain viable it is important that suprasystems maintain structural control over subsystems. This must be done, however, by allowing the component subsystems to retain their identity and to present profitable reciprocal exchanges with each other and with the larger system. If any of these elements is ignored, internal conflict will occur.

Assessing feedback loop operations in suprasystems, systems, and subsystems is a more difficult task. Differentiating which ecologic variables of influence affect only one level of system as opposed to those variables which affect all three levels can be confusing. The same cultural, spatial, political, and economic variables can affect all levels of systems. The effect on any one system will depend on the unique characteristics of that system. For example, inflation and a depressed economy (economic variables) present in the early 1970s influence all societal, family, and individual systems. The impact of this variable differs according to the unique properties of any one of the identified systems. In general sociologic, physiologic, and psychologic variables are most often single-system influences. Political, economic, and cultural variables most often affect several levels of systems.

Assessment of the focal system then includes collecting data from at least three levels of system hierarchy, noting tensions found within each system, and differentiating variables which influence all levels of system functions or only one level.

How then are all these data collected and who should do the collecting? Consider the last part of this question first. The complexity of collecting data on the mezzolevel suggests that a group of assessors is more feasible than going it alone. The same implication is true for planning, implementing, and evaluating at the mezzolevel. The scope of such data collection also indicates that the assessing group be composed of members with a variety of skills and preparation background. The movement to an interdisciplinary approach to health care is in part prompted by these considerations.

The assessment team needs to look at its role in relation to the systems under study. It is necessary to be in the position of having access to relevant data, to have mobility within the designated system, and to disturb the system only minimally during the data-collection process. There are advantages to being a part of the system under study just as there are advantages to being an outsider looking in. Disadvantages are also inherent in each position. Being a component in the system under study necessitates continual self-evaluation of responses to the events under study. Mobility within the system and access to data concerning internal system processes are advantages held by those who are a part of the system. They not only closely observe system processes but are also participants. Less obvious meanings of events are more easily searched out through dialogue with other components. The difficulty in such a position is being able to make observations which disturb the system as little as possible. Changed behavior demonstrated through data collection activities can cause role perception changes that may interfere with objectivity and ability to gather additional data. It is wise for participant assessors carefully to define and clarify their role with other system members. Such definition needs to include prerogatives, limitations, and responsibilities. The "sneak attack" approach may work well for spies but does little to facilitate open relationships among health professionals.

Assessors who are not members of the system under study have the advantage of distance and can look at the complex activities of group roles and comprehensive interaction patterns among components. The outside observers can be considered ecologic variables of influence entering through the feedback loop. As such, valuable data on feedback loop function can be collected. Because entrance into the system is necessary for this work, the system must be open enough to allow the assessors entrance. Closed systems present few data other than their methods for closing out variables of influence. Once more, it is essential that the assessor team meet with the involved members of the system to define and clarify roles. Such meetings facilitate the adequacy of observations and help ensure presentation of valid information.

It is easy to see that the assessor team working within a system provides the most complete information about internal system functions while the outside assessor team gathers the most useful data on feedback loop functions. Elements of both facilitate data collection on overall systems functions. Recent moves toward coalition groups of both consumers and professionals studying community health needs takes these principles into account and, it is hoped, will provide health planners with the most complete data base possible. It is our belief that not only must an assessor team represent a variety of professional disciplines but also it must include consumer members. All members must perceive and react to one another as colleagues.

Let us return to the first question of how all the data are collected. The methods employed in data collection can be categorized by the resultant data—either subjective or objective. Subjective data comprise information collected and processed which can be "biased" by our life experiences. The methodology is usually not systematic, the validity and reliability are dubious, and the accumulated data may not "hold up" under the scrutiny of a scientific critique. The methods often emerge from "hunches," "gut-feelings," and "intuition." It is often this jumble of subjective data, however, that leads to further scientific inquiry and uncovers relationships or needs that have gone unrecognized. R. Buckminster Fuller[3] explores the subjective experience of intuition. He does not put it down as nonscientific and unreliable. Rather, he sees it as

the absolute-velocity insistence of the intellect upon the laggingly reflexed brain to call its attention to significance of various special-case, brain-registered, experience relationships. Intuition is intellect coming instantly in at highest speed into dominance over lower speed, lagging brain reflexing.

Such processing of subjective data makes appropriate use of the assessors' time and provides a springboard for further action.

Objective data comprise information collected and processed in a systematic manner utilizing methods of collection that are valid, reliable, and allow for as little bias as possible. Another characteristic of the data is their multiple use, either across program areas or for comparative studies.

DATA COLLECTION

Subjective Data

Sight, sound, smell, taste, and touch all have their place in assessing a community and its health. To a large extent, the unique characteristics that give a community its identity can be revealed through full use of the senses.

Sight. What do you see when you walk or drive through the community? Are your boundaries a neighborhood, a larger not so politically defined geographic area, or a functional community such as a college campus or an industrial complex? Is it urban or rural? What spatial characteristics can you see? Is this a city on a mountaintop or is it tucked into a crescent around a seashore or river bend? Are the trees tall pine, massive oak, or a stunted mixture of aspen or birch? Is the earth rich loam, blue and red clay, yellow sand, or rocky, or is it leached of minerals and nutrients through ineffective land management? Take the blinders off—look around and describe what you see. With your senses sharpened, you can observe and describe the area in the context of structures. What types of housing, single-dwelling versus multi-dwelling; construction, brick or wood; upkeep, in good repair versus condemned? Do street names and building construction reveal any unique characteristics? For example, an individual placed in New Orleans without knowing the name of the city could suspect the French influence from the street names and architecture. He would not need to be informed of the city's historical background to "see" its European character. Look for the services available to residents of the area. Are there grocery stores, drug stores, medical and dental facilities in the immediate area? Where and what type of churches and schools? Can you ride a bus or get other transportation around, in, and out of the area? Are children playing in the street or are they involved in a planned recreational program? What kinds of animals do you see? Are they finely bred show animals, Heinz 57 variety, or large and fierce and used for protection rather than companionship? What indicators are there of types and availability of water, electricity, and waste disposal services? The small hut with a quarter moon cut in the upper section of the door *could* be something other than a tool shed.

Sound. In some communities, the casual visitor is unable to sort the cacophony that assails the ears from all directions. In other communities, the silence is deafening. Take a trip through the area during the evening as well as during the day. Some areas come "alive" in the late evening hours when children and parents are at home. Listen, really listen, to the sound! By differentiation of sound you can discern the type of traffic patterns, condition of streets, music preferences, and tastes as determined by sounds from radios and stereos, children at play, people working, presence of animal or bird life. In the midst of an urban neighborhood, the sound of a rooster crowing is startling but has meaning when coupled with other pieces of data about economic and cultural influences within the area. Listening will give clues to cultural patterns, language variations, and interpersonal relationship patterns.

Taste. Oh, yes! Taste is important too. What are the food preferences of this community? A meal at a neighborhood restaurant or tavern can reveal much about food preferences and culture of the residents. Are there tacos, neckbones, or roast beef on the menu? A visit to a neighborhood grocery can also indicate food preferences and provide information on prices, which may vary widely from other parts of the community. Attend a church bazaar or neighborhood bake sale. Such a sale in Sault Ste. Marie, Ontario, Canada, which includes Italian bake goods like cream-filled cake peaches and anisette-flavored cookies, alerts the assessor to the unexpected population of Italian immigrants.

Smell. Don't hesitate to take in a big whiff. Sniffing is for the timid. Identify the types of industry. A paper mill smells a lot different from a tool and die company. Does the community use coal or wood as a heating/cooking fuel? Can you identify sources of noxious odors? Is an odor that of waste rotting, an open drainage ditch, or a meal being prepared with unfamiliar foods? Are there sweet smells of jasmine and lilac from the yards or city plantings along the streets? Smell can help identify prevailing wind direction as well as provide specific information on the community's life style.

Touch. Touch in this context is the overall impression—the feeling surrounding the experience of being there. Is this a place of warmth or of hostility? Does it invite the observer to reach out and feel a part of the environment—an iron fence, glossy petaled flowers, white birch trees? Does the community feel closed to outsiders to the extent the observer hurries through the task of data-collection?

In your tour of the identified community, what did you discover about the people? What did they look like? What physical characteristics appeared to predominate?

Subjective data-collection done through observations and casual conversations in laundromats, beauty shops, barbershops, and taxicabs; through eavesdropping in grocery stores and restaurants; through strolling along cracked sidewalks and dusty paths; through smelling local smells and eating local foods—these all provide clues about the community. Putting these observations into the surrounding time frame adds another dimension to the data. Were your observations made at night, morning, fall, or spring? For example, the desert of south Texas is drab in summer and winter but in spring its dull browns are transformed into lawns of pinks, blues, and yellows.

Data obtained by observation provide clues about the community and its problems. Processing of such data offers the opportunity for the formulation of hypotheses that are amenable to further study through the application of more objective methods.

Objective Data

The collection of objective data is necessary to complete the assessment of the community and its health. There are three general areas for objective data collection: spatial, demographic, and resources. Much of this baseline information is readily available through national, state, or local sources which compile such data according to census tracts.

Spatial Data

These comprise information directly related to the geophysical characteristics of a community. Geophysical characteristics are those qualities or properties that shape boundaries or act as natural barriers to growth or utilization of services. The visible types of spatial barriers are rivers, mountains, and limited-access highways. An example of the influence of such a barrier occurred in the midwest, where a river historically divided a city into distinct geographic communities. Later, with the building of several bridges, businesses and services relocated from one of the communities into another, and thus one piece of the city became isolated. Although services and merchandise were accessible by bridges, the "old-timers" still do not cross the river. Highway construction, especially the interstate networks, have displaced persons from familiar neighborhoods. The highway acts as a barrier to friends, churches, and acceptable services. The phrase, "You can't get there from here" takes on new meaning to neighborhoods divided by a limited-access highway. Nonvisible geographic boundaries are those imposed by political structures. State and county boundaries are not painted in red but do present barriers. Though such barriers are invisible, their impact is felt through the educational system and protective services such as police and fire departments. These geographic boundaries have direct effect on assessibility to any health care delivery system.

Other objective geophysical data that have impact on the community and its health are altitude, temperature, and humidity. These factors are "felt" as well as measured. Routine throat-swabbing screening programs to find streptococcus are relatively rare in most parts of this country. Because of the incidence of such infections in the high plains regions of Wyoming, however, such screening programs are carried out weekly in the area's schools. The dry, windy, cold climate of these plains, which are located at an altitude over 6000 feet, predisposes an entire population to respiratory infections. These geophysical data are also important in forecasting potential disasters. Tornado alert programs and flood control plans develop in areas where tornadoes and floods regularly occur. Areas subject to abrupt weather changes are vulnerable to temperature inversions. Such inversions have been held responsible for deaths, as in the London Fog Epidemic and the Donora Fog in Pennyslvania. The Pollution Index was devised to provide the public with information needed to protect those with respiratory difficulties from unnecessary exposure.

Demographic Data

These comprise information describing an identified population. A numbering of the focal community with regard to demographic characteristics is essential for appropriate planning. Baseline data for characteristics such as age, sex, and marital status provide information of special relevance. They are known to be determinants of processes affecting other population characteristics. Birth rates, migration patterns, mortality issues are examples of population characteristics affected by the base characteristics of age, sex, and marital status. Sociologic data which reflect and influence socioeconomic structures are included in information about occupation and income, social and ethnic background, religion, and nationality. The last major group of characteristics has to do with those that reflect population quality. Literacy rate, level and kind of education, morbidity-mortality statistics, levels of immunizations among children are examples of such data.

In each of these major categories of population characteristics, there is also a need to collect data that are labeled deviant and to identify the system's method of dealing with this deviancy. Deviancy is a social definition that includes a description of the deviant's behavior, an evaluation of what the behavior means, and a prescription of how the deviant will act in any given situation. Deviance is in the eye of the beholder, so for deviance to become a social fact several sequential steps must occur. First an act, person, situation, or event must be perceived as a departure from social norms. For example, in the first category of demographic data collection, old age in a youth-oriented society is a departure from the social norm. Being female in a male-directed world, being single in a couples' environment, and other such situations meet the first criteria for the deviant label. The second step involves categorizing the perception into a convenient stereotyped slot. An old man seen scratching himself can be categorized as a "dirty old man" even though he thought he was unobserved in his efforts to become more comfortable. Third, the categorized perception must be reported to others, and fourth, these others must accept this definition of the situation. The last step involves eliciting response from the subject that confirms the definition. For example, a young girl playing in the park observes the elderly man scratching himself. She remembers a warning from her mother about people who "play with themselves" and runs to her mother to report this old man who was messing around in the park. Her mother accepts the story and goes with her daughter to the park. She watches the elderly man closely for signs of deviant behavior and after an hour she verifies her fears when she sees him go into the bushes and relieve himself against a tree. The old man's deviance has become a social fact even though he has been unaware of the process. For the label to stick to him over time, he would have to become aware that others viewed him as a "dirty old man," accept their estimate of him, and seek the company of other "dirty old men."

Such a labeling process can occur in any of the other population characteristic categories. We all are aware of the labeling process connected with religious and ethnic groups as well as socioeconomic status. Simply put, quality of population is determined through what is left over after exclusion of the deviant groups. Of concern to the assessor then is not identification of specific deviants but rather how the group being considered defines deviance. As deviance is a matter of social definition, it is important to recognize that it is this definition process that often produces the deviant act.

Resource Identification

Here, the listing of existing facilities (official and nonofficial), manpower, and money sources comes into play. Who is paying for what, and where? The yellow pages in a local phone book can help provide these initial data. Some communities publish a community resources guide. It is important to learn more than just what services are provided. Money sources, who are the members on the Board of Directors, as well as who is available to provide the service, are data necessary for effective planning. Are there adequate support systems available such as water, waste treatment, electricity, and transportation? What other resources are available to the population for their specific identity needs? School, youth groups, churches, senior citizens groups, etc., contribute to these resources. Location and avenues of access are relevant resource variables.

Objective data collected on spatial, demographic, and resource characteristics describe attributes of system components. Important to this data collection is information concerning the relationships among these components and their attributes. Many of these system processes, as described earlier in this chapter, can be identified by answering these following questions: How are the human needs of the population being met? How is deviance from social norms defined and dealt with? What means are available to individuals for development of their identities?

What tools are available to facilitate collection of objective data related to spatial, demographic, and resource influences? City Planners' offices are excellent sources of information on geophysical data. Most large metropolitan areas and many smaller urban areas have such an office. Some cities use the title City Manager rather than City Planner. Most such offices are affiliated with the Mayor's office but are politically independent. When this source is unavailable, the Chamber of Commerce can often provide such information. County Extension Offices are excellent resources for those in rural areas. Conservation offices also compile this type of information.

Most demographic and resource data can be obtained from existing data. Publications by Bureau of the Census, Public Health Service, and other specific HEW offices can provide a great deal of relevant information. The Bureau of Labor Studies, American Hospital Association, American Nurses Association, and other professional organizations can provide manpower information. (See Appendix D, page 171.)

Where local statistics are unavailable, the use of specific surveys should be considered. This is one of the simplest forms of data collection and among the easiest to administer. Questions should be obvious, requiring little interpretation. It is useful to ask for the same information in two different ways when looking for reliable data. For example, the questionnaire can at one point ask the number of years a couple has been married and later ask the date of the marriage. Development of questionnaire needs to consider the degree of structure which will surround the question itself and its response. A straightforward question can require one answer containing a fact: How old are you? *18* years of age. Other questions can elicit no "one" correct answer. What do you think about the resources; shopping facilities, health services, school, etc., offered to you in this community? This last question could be made more structured by specifying each resource and limiting the response to a checklist:

I do not like the shopping facilities _____
If the above answer is selected, make a statement telling specifically what you do not like.
Statement: _____

I do like the shopping facilities _____
If the above answer is selected, make a statement telling specifically what you do like.
Statement: _____

I have no knowledge of the shopping facilities offered to me in this community _____
If the above answer is selected, check one of the following:
I do no shopping _____
I am new to this community _____

Other reasons I have no knowledge of shopping facilities in this community.
Statement: _____

In summary, the objective data-collection choice is evaluated by its validity, reliability, sensitivity, and meaningfulness. Are you able to gather data relevant to the problem? Are the data precise? Would two interviewers asking the same questions of the same subject get the same answers? Were the results consistent when the same question was repeated? Did the results show significant differences among those questioned when such data were being sought? Did the data provide the necessary information? Questionnaires are good for getting factual information about large numbers of prople. They are less effective in gathering data on emotionally-laden issues.

ANALYSIS AND PRESENTATION OF DATA

The data from both sources of collection are gathered together, shuffled, sorted, scrambled, and spit out in composite form. This process described above is the analysis and presentation of data. The resultant composite will provide a community profile. Community profiles, depending on the size and structure of the community, can be presented in a variety of ways. The most simple is the one- to two-page document as provided by most city offices of the Chamber of Commerce (Fig. 2). This document gives the impression of an active, growing, involved Southern city. Health status composite data are available in the City-County Book provided by the Census Bureau.

But more important than the unrelated facts discovered in statistical printouts are the relationships between factors in the community profile and items provided by the health status data. Which groups within the community are the most affected, which areas of the city are the most susceptible to illness, what are the attitudes and values of the population to health/illness, and what are the patterns of utilization of service? Without some knowledge of such relationships, further planning and action can result in programs, manpower, and money being misplaced. Discovery of these relationships is the field of epidemiology.

Epidemiology

The classic definition of epidemiology is the study of the distribution and determinants of disease in human populations. As communities and disease patterns have changed, however, epidemiology has expanded to include health as well as illness or disease control. In the beginning, epidemiology was

COLUMBUS (Lowndes County)

MISSISSIPPI COMMUNITY DATA

Date compiled __January, 1975__

LOCATION

Distance to:

Birmingham, Ala. _____ 117 miles

Jackson, Miss. _____ 147 miles

Memphis, Tenn. _____ 166 'miles

NOTE: Distance computed between central business districts.

POPULATION

	1970	1960	1950
County	49,700	46,639	37,852
Percent nonwhite	32.9 %	38.1 %	48.6 %
City	25,795	24,771	17,172
Percent nonwhite	37.9 %	40.2 %	46.4 %

Estimated population within 30 miles __150,000__

TRANSPORTATION

Highways serving area:

U. S. primary __82, 45__ Interstate _____

State primary __69, 12, 50, 373__

Railroads: __Illinois Central Gulf, Southern__
__St. Louis-San Francisco__

Piggyback ramp __Yes__

Bus service: Intracity __City Bus Line__

Intercity __Continental Trailways__
__Greyhound__

Motor freight carriers:

Name	Terminal Facilities
Campbell '66 Express	X
Deaton	X
Golden Triangle Freight Lines	
Merchants Truck Line	
Mississippi Freight Lines	X
Pic-Walsh Freight	
Roadway Express	X
Robinson Truck Lines	X
Shippers Express	

Air service:

Nearest commercial airport __Golden Triangle Regional__

Distance __13__ miles Runway length __6,500__ feet

Airlines __Southern__

Daily flights __8__

Other airports and distance __Columbus-Lowndes__
__County__

Waterways: Community on navigable waterway __No__

Nearest port __Tuscaloosa, Ala.__

Distance __60__ miles Channel depth __9__ feet

Parcel service: REA Express, United Parcel,
Continental Trailways, Southern Airways

FIG. 2: Community profile.

75

FINANCIAL INSTITUTIONS

Banks: Number   Total Assets

  Local   3 $ 187,472,200

  Branch bank from another city _____ $ _____

Savings and loan associations: Number   Total Assets

  Local   2 $ 74,062,830

  Branch from another city   1 $ 174,000,000

COMMUNICATIONS

	Daily	Weekly
Newspapers Commercial Dispatch	X	
Triangle Advertiser		X

Radio stations: AM ___3___ FM ___1___

Television stations: Local ___2___ Reception _____

Cable television _Yes_ Reception (with cable) __10__

Telephone service __South Central Bell__

Telegraph service (Western Union) _____Yes_____

Post office class ___1___

GOVERNMENT

Municipal government:

  Type _____Mayor-Council_____

  Zoning regulations _Yes_ Planning commission __Yes__

Police department:

  Personnel ___58___ Patrol cars ___28___

Fire department:

  Personnel: Full-time ___58___ Volunteer _____

  Equipment __7 Pump Trucks, 1 Service Aerial__

  Fire insurance rating ___6th___ class

  Service provided to industrial areas beyond corporate limits
  __Full coverage at $25 per call__

UTILITIES AND SERVICES

Electricity: Supplier(s) _Tennessee Valley Authority_

  Distributor(s) _City of Columbus_

Water:

  Utility name _____ City

  Source(s) _____ Luxapalila River

  Plant capacity _____5,000,000_____ gpd

  Average daily consumption __4,000,000__ gpd

  Peak consumption __6,000,000__ gpd

  Storage capacity __1,600,000__ gallons

Natural gas:

  Supplier(s) __Southern Natural Gas__

  Distributor(s) _Mississippi Valley Gas_

Fuel oils: Available from _____Local_____

LP gas: Available from _____Local_____

Sanitary sewer _Yes_ Percent of community covered __98__ %

Treatment plant type _____ Lagoon

  Capacity _____5_____ million gallons/day

  Present load __97__ %

Storm sewer _Yes_ Percent of community covered _80_ %

Method of garbage disposal __Sanitary Landfill__

RECREATION

Art museum	1	Parks: Local	9
Auditorium/coliseum	1	State, federal	1
Auto race track	1	Skating rink	1
Amateur theatre	1	Swimming pool	5
Ball field	20	Tennis court	20
Bowling lanes	1	YMCA	2
Country club	2	YWCA	
Golf course	2	Other: Annual Spring	
Indoor movie	5		Pilgrimage

FIG. 2. (cont.)

76

LABOR FORCE (County Data)

Estimated labor force: __21,000__ (Dec., 1974)

Males __11,420__ Females __9,580__

Manufacturing jobs as percent of employment __27__ % (1975)

Local high school graduates __600__ (1974)

EDUCATION

Public schools:

	Number	Teachers	Enrollment
Elementary	11	214	4,139
Secondary	4	178	3,488

Private, parochial schools: __3__ __1,021__

NOTE: Public school data by school district. Private and parochial school data by county.

Vo-tech centers:	Enrollment
Golden Triangle Vo-Tech Center	1,903
Lowndes County Vocational Complex	143

Colleges:	Enrollment
Miss. Univ. for Women	2,956
Miss. State Univ. (16 mi.)	10,000
Mary Holmes Jr. Coll. (18 mi.)	325

Public library __1__ Total volumes __60,039__

HEALTH CARE

Hospitals __2__ Beds __199__

Doctors __35__ Dentists __14__
Orthodontists - 2   Oral Surgeon - 1

CHURCHES, SYNAGOGUES

Protestant __60__ Catholic __1__ Jewish __1__

HOTELS, MOTELS

Hotels, Motels __16__ Total rooms __647__

TAXES

(Tax year 19 __74__)

Manufacturer's property taxes (excludes inventory):

Area	Rate/$1,000	Assessment Ratio	Effective Rate/$1,000 of Actual Value Inside City	Outside City
City	$ 19.80	40 %	$ 7.92	$
County	$ 35.60	25 %	$ 8.90	$ 8.90
School	$ 21.30	40 %	$ 8.52	$
State	$ 4.00	25 %	$ 1.00	$ 1.00
Other	$ 33.25	25 %	$	$ 8.31

Total effective rate/$1,000
of actual value $ 26.34 $ 18.21

Local nonproperty taxes:

City:	Retail sales	No
	Income (wage)	No
County:	Retail sales	No
	Income (wage)	No
State:	Retail	5 %
	Individual income	
	Minimum rate	3 %
	Maximum rate	4 %
	Corporate income	3-4 %
	Corporate franchise	0.25 %
	Intangibles	No %
	Gasoline	9 ¢/gal.

CLIMATE

Temperature: Monthly average – January __42.7__ °
– July __81__ °

Rainfall: Annual average __56__ inches

Snowfall: Annual average __0.3__ inches

Prevailing winds __N-8 Knots__

Average elevation __191__ feet (mean sea level)

FIG. 2 (cont.)

77

MAJOR MANUFACTURING EMPLOYERS

Name	Product or Service	Number of Employees Male	Female	Union
American Bosch Electrical Products	Auto Elec. Equip.	569	603	IUE
Seminole Manufacturing Company	Men's & Boys' Slacks	124	371	
Beneke Corporation	Bathroom Seats	400	100	CJA
Johnston-Tombigbee Furn. Mfg. Co.	Bedroom Furniture	292	110	CJA
General Tire & Rubber Company	Vinyl Wall Coverings	620		UPCL
Airline Manufacturing Company	Upholstered Furn. Parts	201	17	CJA
Mitchell Engineering Company	Prefabricated Steel Bldgs.	351		
Hooker Chemical Corporation	Sodium Chlorate Phosphorus Pentasulfide	102	9	CCAW
Blue Bell, Inc.	Men's Jeans	10	153	
Humboldt Products, Inc.	Household Textile Pdts.	50	100	

AVAILABLE INDUSTRIAL PROPERTIES

Name	Acreage Total	Available	R&D Center Site File Code
C & G Property	77	38	44001
Golden Triangle Industrial Park	960	750	44003
Industrial Park South	190	14	44002
Smith Property	91	91	44004
Webb Property	116	116	44006

REMARKS

FOR ADDITIONAL INFORMATION, CONTACT:

Name Jesse N. Moore, Executive Director

Organization Columbus-Lowndes Industrial Foundation

Address P. C. Box 386

Columbus, Mississippi 39701

Phone (601)328-8369

FIG. 2 (cont.)

limited to the study of epidemics and/or the exotic diseases, but today we find increasing emphasis being placed on the chronic debilitating illnesses and on healthy populations to discover reasons and relationships for health. The hope is that such epidemiologic studies will uncover relationships which, if dealt with more effectively, would prevent occurrence of chronic illness. Arthritis is one such example where life stresses are being scrutinized as possible precipitators of disease.

By definition, epidemiology has two components: the study of the distribution of disease or *descriptive epidemiology*, and the study of the determinants of disease or *analytic epidemiology*. Descriptive epidemiology provides a profile of the defined population. Within limitations, the demographer can construct a picture of the population affected by an identified entity. In systems terms, descriptive epidemiology provides data about component attributes such as who they are, where they live, and what they are like. This use of epidemiology provides the base for study of internal system stress. But that is only half of epidemiology; the other half is the study of determinants of disease or analytic epidemiology. Analytic epidemiology proceeds from the profile, adds or subtracts known or suspected variables, and attempts to deduce and identify the impact of the relationships of the variables on the population. Or in systems language, analytic epidemiology is concerned with identification of variables influencing the system through the feedback loop and observes the effect these variables have on the previously described system components. If such a use of analytic epidemiology brings to mind the Black Box concept, it is because early systems theorists described systems through such manipulation of the variables of influence. The Black Box phenomenon describes the theorist's experience of being able to identify inputs into a system and see the resultant outputs without being able to identify specifically what went on within the system to bring about the outcome.

Dr. Kenneth Newell defined epidemiology to include the following characteristics:

1. *The application of the Scientific Method to Medicine* including the clarifying of terms and accurate definition of the characteristics studied (development of a language).*
2. *The attempt to consider an individual as a whole and to explain or describe all the factors which may have influenced him in reaching his present state.*
3. *An appreciation that an individual never stands alone but is a member of a group. This should be coupled with the conviction that many individual characteristics may have a significance to individuals which is meaningless except when studied within a group.*[6]

*He used the term broadly to include health as well as illness.

It was his further belief that while other health professionals do fulfill some of these objectives, it was the epidemiologist who fulfilled all three with the primary focus or concentration on the group aspects of health and disease.

The relationship of epidemiology to the health management of community should not be underrated. The epidemiologist provides the health planner with three types of data: (1) clarification of specified relationships, (2) explanation of local events, and (3) assistance in the provision of health services.

Thus, at the macro- or suprasystem level, the epidemiologist is concerned with discovering the patterns and variables affecting a condition in the total population. From these studies, transfer of known data can occur that will explain local or mezzo/micro-level occurrences. New investigations at the micro/mezzo level most often occur when an unanticipated event occurs, for example, food poisoning at a church picnic or an accident involving industrial toxicology. Sometimes investigations occur at the local level when an event varies to the extreme, making the event unusual. The study of the asthma epidemics in New Orleans is such an example. Epidemiologists assist in the provision of health services by defining specific populations in need. Defining populations in need can provide a planning group with valuable data as to resources, both physical and manpower, required to meet adequately the health needs of the population. As the family is the smallest social unit, so the family is also the smallest population unit for epidemiologic study. From family studies, we have gained knowledge of inherited characteristics that affect disease as well as relationships in the dispersal of disease.

Epidemiology is an applied science. Epidemiology in action demonstrates the application of the synthesized knowledge gained from biologic and social sciences. Thus, epidemiologists are "doers." What is done behind the desk is important, but will never replace completely the on-site, first-hand observations. Snow[8] would never have removed the handle from the Broad Street pump, in London, if he had not seen the pump and its handle and recognized the possible significance of his observation.

The *descriptive epidemiologist* is concerned with the "wider view." The methods are such that enable him to describe the defined population and show "what's happening." The results of descriptive studies can be utilized in a variety of applications and situations. Descriptive studies can produce data relevant to disease prevalence and if the surveillance is continued, data can be produced related to disease incidence. Data can be collected that are pertinent for usage by those responsible for planning health services. Furthermore, descriptive studies are useful in describing the natural history of disease classification. Essentially, the descriptive epidemiologist is concerned with three factors: person, place, and time. Because he is describing what is occurring, he does have the freedom to define his terms to his satisfaction.

This freedom has placed a barrier to comprehensive usage; the barrier that lack of standardization of terms builds. Descriptive epidemiology makes no attempt to manipulate the variables or affect change. Surveys, surveillance, or other types of studies are conducted in a manner that will disturb "normal" activity as little as possible. In describing a community, the total community may be chosen for observation, or if size precludes a total sampling, the community may be described on the basis of a representative sample. By whatever method utilized, the data are collected and analyzed to discover what is happening. The descriptive epidemiologist needs only a knowledge of the population and sharpened powers of observation to perform a service. It is from these descriptive studies that problems are defined and clues provided as to possible causative factors or determinants of health or illness. Not only will needs be identified but populations at risk will be delineated. From these possible relationships, further studies are designed that fall within the realm of analytic epidemiology.

The *analytic epidemiologist* wears the two heads of Janus: one facing the past, the other the future. He combines the knowledge and observations of the descriptive epidemiologist with more sophisticated statistical tools for the testing of the hypothesis derived from the earlier descriptive studies. He compares, contrasts, and finds answers. If the descriptive epidemiologist asks who and what, the analytic epidemiologist asks and answers the why.

There are primarily two approaches to the analysis of epidemiologic data. The first of these is a comparison of the present with the past. This "backward looking" approach is known by either the retrospective or case-history approach. Retrospective studies start with the condition and work backward to discover variables or causative factors. This type of analysis is relatively inexpensive, and results are obtainable in a shorter period of time. Their worth, however, is dependent upon the availability and completeness of records. The advantage of this type of analysis is the discovery of associations or relationships. Thus, it was through retrospective studies that the relationship of birth defects to rubella and lung cancer to smoking were elucidated. The formulation of the causative links or relationships are the first step in the resolution of health problems. As much as is gained through retrospective studies, they do not provide data relevant to risk or the degree to which the identified variables influence the problem.

Thus, we are lead to further in-depth studies that can determine risk or the relative importance of the variable. These studies are known as prospective or cohort studies. Prospective studies begin with the condition and the identified variables and look forward to discover what happens, to whom, how often, and how much. A designated population is selected and followed over time to test the defined hypothesis. These studies are more expensive in terms of manpower, money, and material but over time have proved their worth. Fox et al[2] have categorized four key features of sound prospective studies. These features are (1) a choice of study populations (exposed and nonexposed) that will reflect real differences; (2) the use of appropriate sampling methods in order for results to be generalized; (3) the availability of specific diagnostic methods; and (4) means (money, facilities, etc.) for continued follow-up study.

How does all this gathering of data, subjective and objective, fit together in a real circumstance? Consider the following situation which occurred in the late 1950s in Israel. The source of information and narrator of this illustration is a nurse, Beatrice Riss, who was working as the first Community Organizer for Mental Health in the first Community Health Station created in Israel by the Ministry of Health.

Objective Data:

Spatial: *An immigration village for newcomers located about one hour from the nearest large city. Village capacity 120 families.*

Housing consisted of 4-room cement block 1-story houses with average number of people per household 5
Average temperature range annually: 80°-115°
Average rainfall annually: 2.3 in

Demographic: *Number of females 178 Average Age 24*
Number of males 200 Average Age 28
Average number of children per family 3
Birth rate 2.6
Number of married couples 137
Nations of origin Iraq
Average length of time in this country 8 mon
Religion Jewish-Eastern
Number of males employed outside the home 187
Number of females employed outside the home 20
Average income level per family $3000

Resources: *Facilities*
Community Health Station offering Maternal/Child Services
School
Youth Center—in process
Market Place_____
Manpower
List staff from health station and school
Money
Donation of money for Youth Center by Mayor
Other Governmental Support—Donations

Subjective Data

I came to the Mother and Child Station to work with the public health nurse to help her understand more about emotional care of families. The translated name of this station was "a drop of milk" and it was a place where we hoped to give out milk and love. Mothers from the village regularly brought their children to the clinics and would describe in great detail symptoms of illness their children exhibited. Staff members, all European middle class professional people, became alarmed over the reported incidence of vomiting among the infants brought to the center. Mother after mother brought her child in, talked a long time about what the child had eaten prior to vomiting, the look on his face, the amount, the color, the sound he made. The stories varied little from woman to woman. Some were more graphic in their descriptions than others but all talked among themselves about the children's vomiting and all spent lots of time with the examining staff member telling over again and again what had happened.

I thought to myself something psychological is going on here. Teams of others came to draw blood, take urine and stool specimens and do throat cultures. They spent lots of time analyzing food and swabbing dishes. I walked to and around the village and talked to the people I saw. The public health nurse came with me and we made the trip at several different times in the day.

Early in the morning we saw young children and the women hovering around the doorways saying, "Good-by," to the men. It was like mass exodus of the men as they walked down the streets, climbed on buses, and left for work in Tel Aviv. Then, after they left, the place went to sleep. It reminded me of an affectionate dog that yelps and jumps and wiggles as its owners leave and immediately after they are out of sight circles three times and lies down heavily. The women were young and good-looking but at 12-Noon they were still in their dressing gowns and no make-up, no singing, no chatter-chatter of women together. The young children pulled at their mother's skirts and looked around with big eyes. Other children who had bikes rode around wild in the streets. Boys threw rocks at the school and the principal said, "These Jews are not like others who come here. Their children have no discipline." I didn't see many organized group games, they just ran and jumped in rough play.

A youth center was being built for the people by the mayor. His son had died and he wanted to dedicate the facility in his name. The mayor and his immediate advisors had planned it all and had hired professional workers to build it. Each morning some piece of the work done the day before was found damaged. No one could say who would do such a thing.

At evening when it was almost time for the men to come home, the women would begin to dress up. Their only energy left was to look pretty for their husbands. The food they fixed for evening usually involved very simple preparation. They didn't fix national foods anymore that took lots of time to make. Breakfast was fixed for the men only and their dishes sat on the table until late afternoon. Lunch consisted of a yogurt-like drink, a salad, fruit or some other pick-up type food. It was too hot to cook at noon. When the husband got home she said, "Entertain me, I've been alone all day." Only more babies come from his attempts to comfort her.

These families came from Iraq where they had lived in extended family units. Women were accustomed to being surrounded· by their mothers, aunts and sisters. One woman from the camp seemed more able than the others to see some of the reasons family life was breaking down. She like most of the others had little formal education, but had insight into her own feelings. She said, "I'm tired all the time and yet do nothing. I miss my family (meaning her mother and aunts and sisters). I have no interest in life anymore. I don't read or play music. The only time I talk to anyone is at the center when I see my neighbors and the doctor. The only thing of interest that happens all day is when my baby throws up." This woman was interested in sharing with me more about how life was in Iraq and what was missing here. She also was well liked by the other women and could talk easily with them.

The investigation team could discover no biological reason for the children's vomiting. It was as I had thought the first day, a psychological thing. Everyone was asking for attention from everyone else and no one had energy to give. They were depressed and all that untapped energy went to waste. In the older children it went to acting-out and in the young ones it went to vomiting. The entire family was affected. The resources offered to them from the community were not reaching them. We decided to intervene with two target populations: the acting-out teenagers and the mothers. We would do as our name said, give out love as well as milk. [7]

This situation illustrates assessment as an active meaningful activity involving several different people. The woman from the village, the staff from the center, the Community Organizer for Mental Health, the investigative team, the school principal, the mayor, all contributed to the data collection. The "community of solution" can be seen to have both functional and structural elements. It was a village formed because of a common need and a sense of belonging together. It was also an organizational unit complete with geographic boundaries. The recreational resources available within these boundaries were minimal. The time orientation of the women was past-time while the husbands and acting-out children were involved with now-time. Many of the professional staff and their services were moving in future-time. Professional staff moved quickly; the women moved slowly. Several levels of systems were involved in the assessment beyond the individual children who stimulated the initial professional involvement. Subjective data triggered the "intuition" that nothing physically was wrong with the children. Objective data collection confirmed that intuition and facilitated the hypothesis that the mothers were depressed—grieving for a lost life-style and unable to move toward integration in a new one bereft of their extended families. Analysis of the systems involved showed that the family systems had stresses related to role changes, that the resource systems had stresses related to closeness in including consumer input through their feedback loop operations, and that the whole community as a system had tension among its components with the power groups beginning the labeling process relative to the acting-out teenagers and their families.

Use of descriptive and analytic epidemiology helped define the two target groups identified in the example situation presented in this chapter. Each of the components in the family systems was described. Of these, the children and their mothers were presenting "symptoms." The younger children were described as being more dependent in all ways upon their mothers so the mothers became one target group. The older, acting-out children became the other. Within the total community system, the components were described. The families could be dealt with through dealing with the two target populations. The resources and their lack of openness to consumer input became one target system where new variables would be introduced and the effect upon the resource system and upon the larger community system be analyzed. In this case, the youth center became the target system. Through use of some planned change techniques that will be described in the next chapter, a community representative group met with the mayor who was funding the project and with the engineer who was responsible for the building. The consumer group recommended changes based on their perception of their own recreational needs. Plans were changed and members of the community became active in actual work on the center. That night vandalism stopped, as did much of the acting-out behavior of the teenagers.

An energy flow chart is exactly what it says. It is a picture of the energy flowing through a system. As a tool for organizing data, it can be very useful in pinpointing areas of stress and tension. It would be possible for example, to trace the energy flow of the women in the immigration village (Fig. 3).

Such a flow chart would show that for the two major outside meeting places the women only meet at one of them. It also illustrates the relative isolation of each woman and how even neighbors go alone to and from the center and to the market place at different times. An energy flow chart of the older children could show meaningless purposeless activities in small isolated groups. Interaction energy flow charts could illustrate the relative lack of professional involvement outside of the center. The use of the flow chart can feed into both kinds of epidemiologic approaches by organizing data on

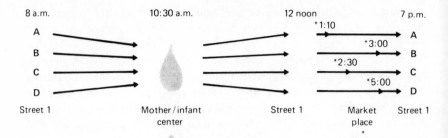

FIG. 3: Assessment through energy flow chart.

characteristics of the population and by tracing the flow of one variable through a system.

Force field analysis and cybernetic analysis are two tools which go together. Both have to do with feedback loop operations and both can facilitate the application of analytic epidemiology. The force field analysis, as described and used by Kurt Lewin,[5] essentially identifies variables of influence which are causing a system to maintain a status quo. It is also a tool that can be used in planned change, as described in Chapter 4. For our example, the women of the village were seen to be unable to change their behavior in this environment. A force field analysis would seek to identify those forces influencing the women to remain depressed and those forces which were influencing them to become less depressed (Fig. 4). This is a

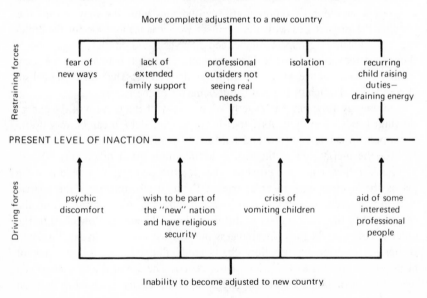

FIG. 4: Assessment through force field analysis.

simple force field analysis that shows only sample restraining forces and driving forces. It demonstrates visually the data gathered which relate to movement or lack of movement within a system. It complements analytic epidemiology by articulating the variables currently influencing a given system. Cybernetic analysis follows this same principle with one important difference. A cybernetic analysis would include a value weight expressed in a number with each variable. For example, on a scale of 10, fear of new ways may be given a value of 7. Psychic discomfort might also be given a value of 7, thus, making equal two variables and maintaining the status quo. Cybernetic analysis provides a number analysis of variables and for complex systems necessitates the use of a computor to analyze accurately all the variables. Social systems research has not become sophisticated to the point where such cybernetic weighting values are available for broad application. Those interested in becoming involved in refining a new tool for use with social systems will find such analysis a wide-open interesting field.

Presentation of Data

Presentation of data is important no matter what tool or tools are used to organize the data. Often, time and money have been lost not because of lack of justification for a proposal or validation of a need, but because the method of data presentation was inappropriate. For this brief discussion, two facts should be kept in mind: who will look at the data, and are the data clear or concise enough to be understood by the viewers?

For the presentation of geophysical data, mapping is probably the method most easily understood. If the services of a cartographer are not available, there are books that are of assistance in preparing displays. Commercial maps of areas of political jurisdiction are usually readily available. From these, overlays can be made that will demonstrate the different variables having impact on a specific geographic region. Maps are indispensable for demonstrating clustering of cases or conditions. Herzog, in his novel, *The Swarm*[4], not only demonstrates the team approach to problem solving, but has included maps that provide additional impact and veracity to his story. Figure 5 is a series of maps that demonstrate how the amateur cartographer can organize data for pictorial presentation. Map 1 demonstrates the location of health care facilities in a metropolitan area, Maps 2 to 5 describe a rural community as to location, land usage, and structural relationships. Map 6 depicts population distribution by identified housing units as to disposition and study results. Map 7 demonstrates the relationship of study results to known conditions throughout the total defined geographic area.

Another method of presenting geophysical data is photography. This is especially useful if an area exhibits unique characteristics. Perhaps the

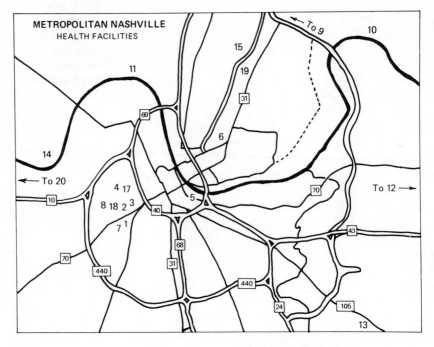

FIG. 5: Demonstration of data presentation (Maps 1 to 7).
Map 1.

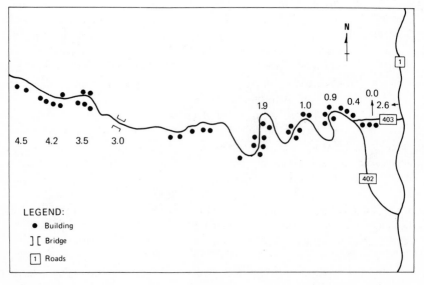

Map 2.

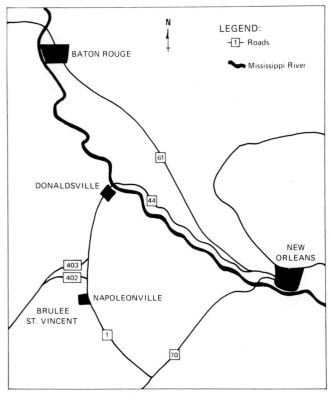

LEGEND:

─[1]─ Roads

◄ Mississippi River

BATON ROUGE

N

DONALDSVILLE

61

44

NEW ORLEANS

403

402

NAPOLEONVILLE

BRULEE
ST. VINCENT

1

70

Map 3.

architecture, the wildlife, or some occupational feature would provide impact on presentation in this manner. Here again, the services of the professional could be useful, but with the new almost error-free cameras the amateur can do much himself in recording what he observes. ·

Another method for the presentation of data is through the manipulation of numbers. There is a variety of ways that numerical data can be presented in tables and graphs of various types. Often the data are such that their sheer weight makes these difficult to handle. Through the use of statistics, data can be manipulated and transformed into numbers that can be handled and understood. Essentially, statistics can be categorized into two types: *descriptive* statistics and *inferential* statistics. Descriptive statistics are those methods used to summarize data. Large masses of data are arranged in such a manner that they are readily understood. The rate is an example of a descriptive statistic. Inferential statistics are those methods used to test hypotheses, or the means whereby a generalized statement can be made after sampling a portion of the population. There are two considerations when reviewing statistical data: (1) Were the representations appropriate to the data?

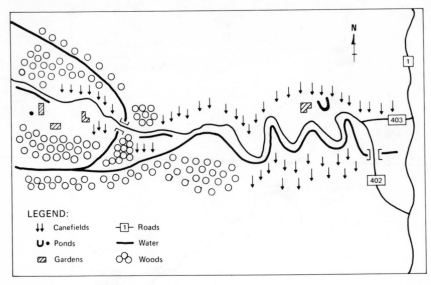

Map 4.

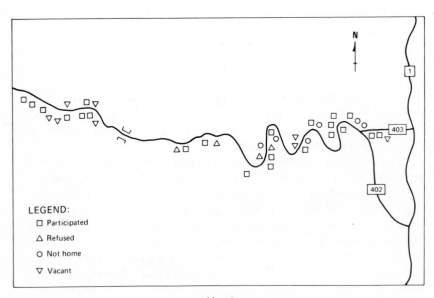

Map 5.

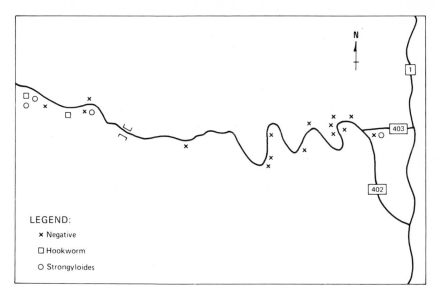

Map 6.

(2) Did the statistic accurately represent the facts? A short summary of common descriptive and inferential statistics has been included in Appendix E.

CONCLUSIONS

The statement of conclusions has two essential components: the identification of need (includes present status as compared to desired status) and the system analysis that enables the predictions used later in the planning phase effectively to implement planned change. These two components are analogous to the diagnosis and prognosis of an individual. The diagnosis identifies the conditions and causes while the prognosis predicts future conditions within certain limitations. By judicious use of the assessment tools discussed here, the collected data should clear up some of the mysteries of the "Black Box" without interfering with the equifinality effect.

Collection and analysis of data will lead to a description of the current health status. It is assumed, here, that the community or agency that has approached this process has developed criteria to describe the desired status, whether it is to reduce maternal mortality to below national levels, or to raise the immunization level of children 0 to 6 years of age to the minimum acceptable levels. Often the initial data collection is only compared with state

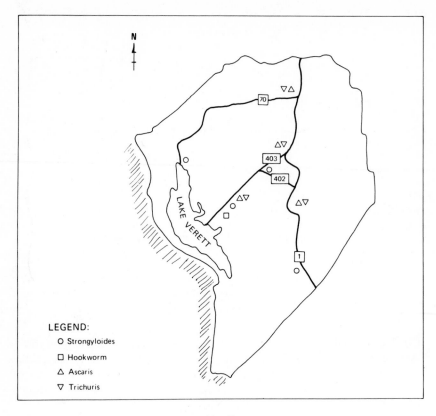

N
↑

LEGEND:
O Strongyloides
□ Hookworm
△ Ascaris
▽ Trichuris

Map 7.

and/or national statistics, and once these gaps or discrepancies are found, the community decides what goal to shoot for.

The systems analysis will reveal the patterns used in the past to meet needs, the resources available, and the factors or variables that will influence the implementation of planned change. (See Appendix B, Cursory Assessment of a Selected System, for an idea on how to organize the gathered data related to a specific system.) Use of energy flow charts can also graphically picture interrelationships among systems that are related to a specific problem.

Another way to organize data is analysis of the ecologic variables of influence related to any specific health need (Fig. 6). Of these variables, the political, economic, socio/cultural, and spatial factors are most likely to influence all systems under assessment.

Political.   What structures are operational that have impact on health? In essence, who makes the decisions? Does the power come through the formal structure or through more informal structures? Change is usually easier if the

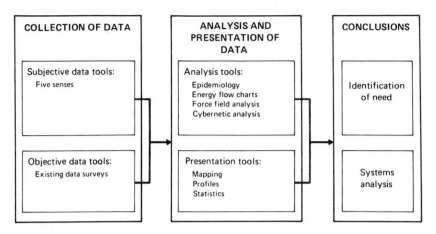

FIG. 6: Assessment model of community health needs.

power structure has been convinced of the need for change. Do these structures have jurisdiction over the area of prime interest? Has the political structure a good history in respect to meeting health needs?

Economic. Is money available now for the enterprise? Is there a mechanism available that can underwrite the cost of a program on a trial basis?

Socio/cultural. What value does the population place on health? How do they view their health status? Do they consider the community a healthful place?

Spatial. In this discussion, the spatial variables refer to existing health care systems. Have the available services been evaluated? Which services are lacking? Are they services the community feels are necessary rather than services that are more fadlike? What are the existing programs? Do any of them meet the defined need? What type of constraints exist in terms of action? Will there be a need for change in the law or procedure to accommodate change? What communication/interaction patterns are present? The current status of coordination and interaction is important. If a network is already present that can be utilized, time, effort, and money can be saved.

What other subsystems of the community link or interface with health care? Health and health care services are only one component of a community, and a component that overlaps into many other subsystems within a community. These interrelationships need to be defined. Including this facet in the analysis will lead to the discovery of relationships that could in turn lead to the development of realistically planned collaborative programs.

Assessment is a complex task which can bog down those wishing to be more effective in the provision of health services. When such assessment is done incompletely, however, the chances for implementing change decrease and the chances for spin-off untoward results increase. Such has been our past history in the health care system and the time has come to be truly professional in our assessment activities and to do the task well. Figure 1 presents a model depicting the assessment steps discussed in this chapter. It is another example of an energy flow chart, in this case, a flow chart of activities included in the assessment process.

REFERENCES

1. Berrien K: General and Social Systems. New Brunswick, N.J., Rutgers University Press, 1968
2. Fox J, Hall C, Elveback L: Epidemiology: Man and Disease. New York, Macmillan, 1970
3. Fuller B: Synergetics: Exploration in the Geometry of Thinking. New York, Macmillan, 1975
4. Herzog A: The Swarm. New York, Simon & Schuster, 1974
5. Lewin K: Quasi-stationary social equilibria and the problem of permanent change. In Bennis W, Benne K, Chin R (eds): The Planning of Change. New York, Holt, Rinehart and Winston, 1961
6. Newell K: Epidemiology. Handout distributed to class. New Orleans, Tulane University School of Public Health, 1966
7. Riss B: Personal communication, 1975
8. Schoenberg B, Mann R, Kurland L: Snow on the water of London. Mayo Clin Proc 49:680-84, September 1974

RECOMMENDED READING

American Public Health Association: Guide to a Community Health Study. New York, APHA, 1961

Bancroft H: Introduction to Biostatistics. New York, Harper, 1965

Barker D: Practical Epidemiology. London, Churchill Livingstone, 1973

Bell D: Twelve modes of predictions: a preliminary sorting of approaches in the social sciences. Daedalus 93(2):847-48, 1964

Bergwall D, Reeves P, Woodside N: Introduction to Health Planning. Washington DC, Information Resources Press, 1974

Conner D: Understanding Your Community, 2nd ed. Ottawa, Development Press, 1969

Garlington W, Shimota H: Statistically Speaking. Springfield, Ill., Thomas, 1964

Greenhood D: Mapping. Chicago, University of Chicago Press, 1964

Herban N: A survey of the Brulee St. Vincent, Assumption Parish, Louisiana, for the prevalence of intestinal parasites. Tulane University School of Public Health, Masters Thesis (unpublished), 1967

Jacob F: The Logic of Life: A History of Heredity. New York, Pantheon, 1974

Jones M: Technology Assessment Methodology: Some Basic Propositions. Washington DC, Mitre Corp., 1971

Klien D: Community Dynamics and Mental Health. New York, Wiley, 1968

Lester M: Every nurse an epidemiologist. Am J Nurs 57:1434-35, 1967

MacMahan B, Pugh T, Ibsen J: Epidemiologic Methods. Boston, Little, Brown, 1960

National Commission on Community Health Services: Health is a Community Affair. Cambridge, Harvard University Press, 1967

Rubington E, Weinburg M: Deviance: The Interactionist Perspective, 2nd ed, New York, MacMillan, 1973

Taylor I, Knowelden J: Principles of Epidemiology, Boston, Little, Brown, 1964

USDHEW: Attribute Sampling Methods for Local Health Departments. Washington DC, U.S. Gov't Ptg. Off., 1965

WHO Expert Committee on Health Statistics: Sampling Methods in Morbidity Surveys and Public Health Investigations. Tech. Rpt. Series No. 336. Geneva, WHO, 1966

4
PLANNING AND IMPLEMENTATION

Although planning and implementation are two distinct steps in the decision-making process, they will be considered together in this chapter. As previously mentioned, all these steps are highly interrelated, with the "plan to do" and the "doing" being very dependent upon one another. Data collected, sorted, shifted, and analyzed; problems identified; and many work hours previously spent in assessment activities have provided the base for planning and implementation. That is just the beginning. Problem identification (assessment) is meaningless without action directed toward solution. Such purposive action is initiated through planning. Planning is an "action" word implying a decision-making process. Such a process includes goal setting, ordering of priorities, seeking routes or methods for goal attainment, and selecting the appropriate avenue for action within the framework of the variables of influence. Planning utilizes the data collected, analyzed, and interpreted during the assessment phase and sets into motion activities designed to harness all available resources into the most appropriate and effective channels for implementation.

Planning and intervention imply concern with a change process. Planning would not be necessary if change naturally occurred as an ordered process. As such is not the case, planning can help cope with and perhaps master the seeming chaos surrounding us. Planning provides the means for responding in an "anticipatory" manner rather than a "handle the crisis only" orientation. It is a future-oriented process in that it lays out an orderly design for reaching a prospective destination. Planning is an attempt to anticipate problems rather than to "solve existing problems."

PLANNING

The characteristics of the planning phase are related to two broad elements: identified needs and solutions (Fig. 1). The planning phase requires people who have the judgment to define priorities, set goals realistically, and define and set objectives into operation; the creativity to develop alternative solutions to problems; the management skills to allocate resources; and the leadership ability and wisdom to implement the appropriate solution. Those involved with the planning/implementation process have been identified as change agents. It is likely they are called by a number of other names since it is common to take the attitude that change is threatening. People associated with planning for change have a reputation of being a "special breed of cat." Ralph Nader is a contemporary example. Typically, such people are described as "being ahead of their time" and often they are not financially rewarded for their efforts. Because of society's current views on change and on those who promote it, it may happen that the broadly based assessment team composed of several professional disciplines and consumers will dwindle to a handful when the time for planning and implementation arrives. It is useful to continue to incorporate as many professionals and consumers as possible into the planning/implementation team and to maintain continuity within the membership. Perhaps as change becomes more accepted as a fact of modern life, those who seek to plan changes more judiciously will be less ostracized. The planning phase is analogous to the plan of treatment or the treatment regimen as ordered by the health professional for an individual or family.

Palmiere,[5] in 1972, outlined three types of health planning, as did Bergwall[2] in 1974. Table 1 presents these types as they relate to systems levels. These types of planning may occur at any system level, but one form is

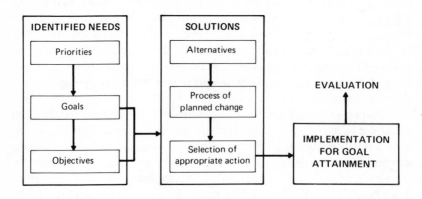

FIG. 1: Planning and implementation model.

TABLE 1. TYPES OF HEALTH PLANNING

FOCAL UNIT	PALMIERE[16]	BERGWALL[5]
Micro/mezzo (Individualized components)	Dispersal	Project
Micro/mezzo (system linkages)	Focused	System
Mezzo/macro (legal responsibility)	Central	Policy

usually more predominant at one level than at another. (1) *Dispersal* or *project planning* has been the most commonly used type of health planning. As our society has become more complex, however, the need for other types of planning has arisen. Each form of planning has its disadvantages. Dispersal planning, with its emphasis on the individualistic role of the components, has often resulted in gaps as well as overlaps. An example is the multiple uncoordinated health services found in any city. (2) *Focused* or *system planning* has resulted in coordinated efforts, but in line with Palmiere's definition of focused planning, this planning comes about because of voluntary association. Because of this voluntary association, there is little or no control over the available resources. Examples of this type of planning are the Comprehensive Health Planning Agencies. (3) *Central* or *policy planning* occurs at governmental levels and uses legislative power to effect change, usually through directing the allocation of resources (usually money). Because of the cultural (attitudes and values) system in which we live, all three types of planning will probably continue to exist for some time.

Characteristics of the planning phase are related to two broad elements: *identified needs* and *proposed solutions*. The characteristics that we are outlining as part of the planning component are still to be considered regardless of the type of planning that is being done. The emphasis on any one element, however, may be greater, depending on the type of planning being undertaken.

Identified Needs

There are three elements to be considered in planning to meet identified needs: ordering of priorities, setting of goals, and determining objectives. The assessment process has provided data relevant to identified needs. Health needs are so interwoven into the fabric of our lives that there is rarely one need but many that interrelate and overlap. These three elements are related yet flow in a sequential manner.

Priorities. In effective planning, the first step toward goal-attainment is setting priorities. Setting priorities can be the hardest task the planning group may tackle. Structural characteristics of the planning/intervention group itself are important considerations for priority-setting. If the group is meant to be a short-term blitz team, its priorities can differ from a team coming out of an established long-term service system. If it is a group which wishes to gain strength but is not yet well supported by the surrounding society system, its priorities will be organized around the group's need to stabilize. Consumer advocate groups have learned the hard way how to accomplish this stabilization. These groups now establish priorities according to issues (identified needs) which are currently visible to the population, to issues which are specific enough to be discussed in action terms, to issues which are possible to win, and to issues which will further build the organization through addition of new members and development of new skills. Success breeds success. A planning/intervention group that has no past history of achievement should beware of risky issues which could endanger their own existence. Short-timers or blitz groups can more easily take on such issues and organize their priorities on that basis.

There are some guidelines other than the group's internal goals which can be used for priority setting. Planning/intervention groups can set priorities as they relate to the following three broad areas of concern: deviation-correction issues, uncontrollable variable impact, and future energy need implications.

Deviation-correction issues come about when something has gone wrong in an existing program and the service system needs to repair the deviation. A planning group working with this area of concern is acting as a deviation-counteracting feedback loop. Priorities for this group will be organized around the need to maintain the system in its original state. For example, a federally funded well-baby program finds its enrollment decreasing. Continued funding depends on maintenance of a minimum census. Priorities will be developed from the need to discover what is happening, why it is happening, and what can be done about it. Another example at a more complex system level is the attempt of society to deal more effectively with deviant groups. Planning/intervention groups dealing with deviations need to be involved with priorities related to returning the situation to previously established norms or to adjusting the norms to accommodate a broader range of deviation.

Uncontrollable variable impact issues have to do with a system's need to deal with an unexpected event over which the system has no control. Planning/intervention groups organized out of this type of concern act to return the system to a steady state. Priorities are ordered from the need to regain equilibrium. Impact of a natural disaster on a community and the activities generated by systems inside and outside the immediate area is an example. On a smaller system level, sudden absence of key personnel for an extended period necessitates setting priorities to return the system to a functional level.

A third broad category of concerns comes out of a system's need to plan for intermediate and *long-range energy demands*. These planning priorities are most often accomplished at a time when all is going well for the system and they are organized in view of future needs. Health care systems are notorious for their lack of planning for future concerns. Under the impetus of the consumer movement and the federal government's concern over rising medical costs, however, planning for future needs is becoming more evident. The National Commission on Community Health Services formed in 1962 is an example of a planning/intervention group organized to deal with future health needs. Another example is a hospital which establishes a planning group to deal with anticipated growth needs and/or the increasing tendency toward use of outpatient services rather than hospitalization. These planning/intervention groups set priorities related to energy demands and anticipated system growth. It is unfortunate that often priorities have been made not on the basis of careful study but either on the basis of current available funding (a type of bandwagoning) or pressure applied from interest groups. It is to be hoped, as the planning process becomes more professional, presentations will be made so the public can better understand the rationale behind the priorities as set by the agency.

The assessment phase can provide data that will assist in making a rational decision as to the ordering of priorities. Examining each identified need or problem in the light of the following will give new insight and knowledge as to how to proceed with planning.

1. What is the magnitude of the problem? It is reasonable that a problem that could cause death or disability would be given a higher rating than a problem of less magnitude or impact.
2. How many people are affected? This characteristic is linked with No. 1 in that it would be expected that a communicable disease like diphtheria would be considered before a condition that is relatively rare.
3. What types of resources are available? In addition to the usual manpower, money, and facilities, consider also the availability of current technology to master the problem. It would be unrealistic to plan for an eradication program if the scientific means were not available for use. Consider too, what is already being done and by whom. Has money recently been made available, are there new breakthroughs in knowledge?
4. What are the prevailing attitudes toward the problem? This calls for an examination of the current attitudes and values relevant to the problem. Which groups are expressing the most interest in the problem?
5. What are legal constraints? Are there restrictions that affect the problem? Are there new laws that state such a problem will be solved?

The International Technical Assistance Programs of the United States government developed an approach to setting priorities; this approach was later refined by Dr. John J. Hanlon.[3] In this process, all needs would be identified and rated against certain criteria and given a numerical score. Hanlon found that through utilization of this process most activities or programs fell within three categories: (1) those with a score of 3 or 4 (the program would be continued); (2) those that have special justifications or reservations; and (3) those that are not justified except in rare instances. Essentially, this tool is a cybernetic analysis of specific variables of influence.

The rating formula has four components with designated ranges of scores as follows:*

Range of Score

0-10	*Component A = size of problem*	
0-20	*Component B = seriousness of problem*	*Public*
0-10	*Component C = effectiveness*	*Health*
0 or 1	*Component D = PEARL*	*Need[3]*

Hanlon gives the numerical computations as follows:

Basic Priority Rating (BPR) $\qquad = \dfrac{(A+B)C}{3}$

Overall Priority Rating (OPR) $\qquad = \dfrac{(A+B)C \times D}{3}$

The maximum product is 300 and the maximum score would be 100. The range of all ratings will be 1 to 100.

Component A.

size of problem is defined as the total number of persons having the problem or being directly affected by the problem expressed in rates per 100,000 population.[11]

The scoring for this component is as follows:

Units per 100,000

Population (Component A)		Score
50,000	or more	10
5,000	to 49,999	8
500	to 4,999	6
50	to 499	4
5	to 49	2
.5 to	4.9	0

*From Hanlon: Public Health: Administration and Practice, 6th ed, Mosby, 1974.

Component B: Seriousness of Problem. Four variables are to be considered in rating the seriousness of the problem: urgency, severity, economic loss, and involvement with people. Each factor is rated on a scale from 0 to 10, but the total composite score cannot be over 20. Thus, if a problem rated 10 on each variable, the score for the component would still be 20. The scoring for this component is as follows:

Factors Comprising Seriousness	Range of Score
Urgency	0-10
Public Concern	
Public Health Concern	
Severity	0-10
Mortality	
Morbidity—degree and duration	
Disability—degree and duration	
Economic loss	0-10
To individuals	
To community	
Involvement of People	0-10
Potential-*number of persons who may acquire problem or be affected by existing problem, and relative degree of involvement.*	
Indirect—*number of persons affected socially, economically, psychologically, etc. and relative degree of involvement.* *	

Component C: Effectiveness. This component implies evaluation and evaluative methods which at present are still being developed. This component is scored in a range of 0 to 10. In addition, it is suggested by Hanlon[11] that values be assigned to the component as follows:

C_1 *for the program or activity as it is now being carried out*
C_2 *for the program or activity as carried out optimally.*

Component D: PEARL. This component is a group of factors that while not directly related to need certainly will determine if the program is to be carried forward. Each variable is assigned a score of 0 or 1. The composite score for component D is the product of the scores of the five variables, either 0 or 1.

P = *Propriety*
E = *Economics*
A = *Acceptability*
R = *Resources*
L = *Legality*[3]

*From Hanlon: Public Health: Administration and Practice, 6th ed, Mosby, 1974.

The original formulas indicate that the Basic Priority Rating could be very high yet the OPR could be 0 because of the impact of the variables rated in PEARL. The rating on PEARL would indicate areas or variables that could benefit from the process of planned change.

The process of placing a value on any one variable could be largely subjective, but objectivity can be built into this exercise through use of scientific control methods. Use of a representative sample of qualified raters, precise operational definitions of terms, delineation of exact rating procedures, utilization of statistical data to guide ratings, and a statistical averaging of scores derived by raters working independently are all ways to increase objectivity and to guarantee the scientific soundness of the results. As discussed in Chapter 3, there is a great need to refine cybernetics as a research tool for social systems. In the twenty years since the priority rating approach was first·developed, its refinement as a tool has progressed to choice and definition of the variables, establishment of usable numerical value ranges, and statement of two formulas defining the relationship among these weighted variables. What remains to be accomplished with this tool is enough applied use under scientific conditions to support its reliability over time. When this scientific research and refinement have been accomplished, this tool may stand as one of the first unchallenged cybernetic tools for social system study.

Let us look at what have been already established as helpful guides for use of an adaptation of this priority-setting tool.

Variable A. This has to do with the size of the problem and has a value range of 10 points (0 to 10). For purposes of assignment priority, the size of a problem is defined as the total number of persons having the problem or being directly affected by the problem expressed in rates per 100,000 population. Should a problem be defined in terms of an area having a large population, the scale can be adapted to apply such as:

POPULATION IN CITY	SCORE
(variable A)	
1,000,000+	10
100,000 to 999,999	8
10,000 to 99,999	6
1,000 to 9,999	4
100 to 999	2
0 to 99	0

Smaller populations can fit the scale if the same ratios are kept constant. The variable would then need to be redefined in terms of a rate per smaller population unit, as example, rate per 100 population.

Variable B. The seriousness of the problem has been defined in terms of four factors: urgency, severity, economic loss, and involvement of people.

This variable has a weight range of 0 to 20. Each factor of seriousness is assigned a range of 0 to 10. It is possible for consideration of only two factors to add up to the maximum score of 20. For Variable B scoring, there will be no change in scoring from the previously stated component B.

Variable C. The effectiveness of existing services has not yet been studied comprehensively, so the factors to be considered have not yet been specifically identified. We suggest that this variable can be considered in terms of systems analysis as discussed in Chapter 3. Those services currently dealing with the problem need to be assessed as systems in terms of internal and external stresses. The value range is 0 to 10 and functions as it does in Variable B, where any two factors considered could add up to the maximum score but 10 does remain the limit even if all factors are rated at very high levels.

FACTORS COMPRISING EFFECTIVENESS	RANGE OF SCORES
(Variable C)	
	0 to 10
Degree of Resistance offered to consumers by the systems concerned with the problem (the higher the rating, the greater the resistance)	
Degree of Nonsummativity among systems concerned with the problem (the higher the rating the lower the degree of interdependence)	0 to 10
Relative openness of involved systems to intake of data about the problem (the higher the rating the more closed the system)	0 to 10
Consistency of the product or output of involved systems with systems statements of purpose (the greater the consistency the lower the rating)	0 to 10

Variable D. PEARL consists of five factors each given a value of either 0 or 1. Since together they represent a product rather than a sum, if any one of them is rated 0, the whole variable D is rated 0.

These variables, considered in their relationship with one another as expressed in the formulas, show that it is possible for a problem to have a high basic priority rating (BPR) but a zero for an overall priority rating (OPR) because of impracticalities at the moment. If this pattern occurs, it is necessary to establish priorities around some intermediate goals. The following is an example of the use of this priority rating scale in an actual situation.

A planning/intervention group assessed the problem of adolescent idiopathic scoliosis in a community of 1,000,000 people. Their assessment included data from literature review supporting a 4% incidence rate among the general population. They also found six service systems currently dealing with some aspect of scoliosis identification and treatment. No comprehensive scoliosis screening program existed in the community. Concerned orthopedic physicians expressed dismay over the number of adolescents referred to them late in the progress of the condition as well as feeling responsibility to participate in a screening program. With other data collected the following picture developed concerning the priority for development of a comprehensive scoliosis screening program.

Variable A—Size of the Problem	Score
Estimate 4% of general population: result, 40,000	8

Variable B—seriousness	Score
Urgency	
Public Concern	3
Community Health Concern	4
	7
Severity	
Mortality	0
Morbidity—degree and duration	5
Disability—degree and duration	7
	12
	(maximum 10)
Economic loss	
To individuals	7
To community	5
	12
	(maximum 10)
Involvement of people	
Potential	4
Indirect	5
	9
	(Total score 20 maximum possible)

Variable C—effectiveness	Score
Degree of resistance	8
Degree of nonsummativity	9
Relative openness	5
Consistency of product and purpose	7
	(Total score 10 maximum possible)

Variable D—PEARL	
Propriety	1
Economics	1
Acceptability	1
Resources	1
Legality	1
	(Total score 1)

$$\text{BPR} = \frac{(8 + 20)10}{3} = \frac{280}{3} = 93.3 \text{ out of 100 maximum possible}$$

$$\text{OPR} = \frac{(8 + 20)10}{3} \times 1 = \frac{280}{3} \times 1 = 93.3$$

The planning/intervention group could have shared the process of determining the priority rating for the potential program with those service systems currently dealing with the problem. A symposium was held and the services involved committed themselves to pursuing the problem further as an united group rather than arbitrarily to develop an entirely new program out of a new agency.

Another such planning/intervention group used the same process to establish the priorities for development of child care facilities for children of adolescents attending junior high and high school. Their data looked like this according to the priority rating scale.

	Score
Variable A—size of problem	7

Variable B—seriousness

Urgency	
Public concern	1
Community health concern	2
Severity	
Mortality	0
Morbidity	3
Disability	2
Economic loss	
To individuals	8
To community	8
Involvement of people	
Potential	4
Indirect	5
	(Total 20 maximum possible)

Variable C—effectiveness

Degree of resistance	10
Degree of nonsummativity	9
Relative openness	0
Consistency of product with purpose	1
	(Total 10 maximum possible)

Variable D—PEARL

Propriety	1
Economics	0
Acceptability	0
Resources	0
Legality	0
	(Total Score 0)

$$\text{BPR} = \frac{(7 + 20)10}{3} = \frac{270}{3} = 90 \text{ out of } 100$$

$$\text{OPR} = \frac{(7 + 20)10}{3} \times 0 = \frac{270}{3} \times 0 = 0$$

Needless to say, the planning/intervention group was unable to establish development of a child care facility within the adolescent educational program. Intermediate problems became the priorities, such as, seeking to provide data to involved systems in such a manner that they would become more open to studying needs of the acting-out adolescent, working to change legislation that currently blocked development of such a facility, seeking to educate the public to needs of adolescent parents, etc.

There may be other priority-setting tools in the developmental stage that will be as useful as this priority-rating scale. It is hoped, however, that planning/intervention groups will make use of this tool in order to refine it for more definitive work in the future.

Goal Setting. Needs have been identified and priorities set; this leads us to the next step—that of goal setting. Each need should have a designated goal to be reached. Goals have value by serving as an overall frame of reference. There are four factors to be considered in goal setting.

1. *Goals are broad statements of desired results.* For example, the goal will be an increase in immunization levels of the population. The desired result is an increase in immunization level. The statement, however, is incomplete.
2. *Goals are time- and space-limiting.* For appropriate utilization of resources, goals should define the time and space limitations. Now our statement would read: The goal will be an increase in immunization level of all children in X-county by 1978. This is better but still incomplete.
3. *Goals are congruent with needs and standards.* Again, for appropriate utilization of resources, goals should be population specific. The assessment defines population of need or at risk. Governments have health standards to be met. The statement now reads: The goal will be an increase in immunization levels of all children in X-county by 1978 between the ages of 0 and 6 years of age.
4. *Goals are realistic targets for purposive action.* Our example is still incomplete. The problem is the lack of quantitative statements to proceed with actions. The phrase "an increase in immunization levels" tells us nothing of the present status or the target we wish to meet. The current status is known from the assessment phase; the ideal may be stated in the standards. What quantitative measure will we be satisfied with? Goals, then, often represent a compromise between the ideal and what can be done with the resources available. So, our completed statement now reads: The goal will be to attain an immunization level of 70 percent (current status 40 percent) in the children 0 to 6 years of age in X-county by 1978.

This gives us a broad statement of what we wish to obtain but not the route to get there.

Why do we need goals? Why can not we just start acting now that needs and priorities have been identified? Goals are helpful for three reasons.

1. *Goals provide a broad base for outgrowth of specific objectives.* We know now from our goal that we have a date for accomplishment of the task, a specific population to be served, and a specific quantitative measure to reach. From these guidelines, specific measurable objectives can be formulated in line with available resources.
2. *Goals provide purpose for expansion or design of services.* Do the present resources have the capabilities to meet the goal? The gaps in service or in structure may become visible so that a realignment is made for better utilization of resources.
3. *Goals assist in reduction of frustrations.* When goals are not defined, personnel may be overwhelmed when facing needs with limited resources. Specific goals provide realistic targets and reduce frustration level.

Objectives. Objectives are the specific avenues by which goals will be met. It is generally accepted that objectives should meet the following criteria:

1. *Be within the scope of the organization.* Does the organization have the authority to initiate and complete the intended action? Does the organization have the resources (manpower, money, material) for the task?
2. *Be congruent with the stated time-frame.*
3. *Be acceptable.* Do the personnel, available and responsible, find the service acceptable? In addition, does the consumer (individual, family, or community) find the service acceptable? Objectives that conflict violently with cultural systems have little chance of success.
4. *Be legal.*
5. *Be measurable.*

One of the most difficult tasks related to writing objectives is meeting the criteria for measurability. Objectives should be expressed in terminology that delineates quantitative as well as qualitative characteristics. One of the tools currently in use to facilitate this specificity is the behavioral objective. Behaviorists have made wide use of this tool in behavior therapy and behavioral learning theory. Essentially, the behavioral objectives define the goals in terms of observable behavior. Such objectives include more than means for achieving goals. They also spell out costs and benefits, reallocation of workloads, and possible consequences of different kinds of action. The planning process includes more than designation of the most appropriate means for given ends. It also seeks to predict the possible consequences, to clarify society's values, and to make people aware of the cost of achieving such changes. To accomplish these tasks, the behavioral objectives are developed in the context of the following characteristics:

1. *Behavioral objectives should be stated in terms of behaviors of the client system rather than in terms of the change agents.* The major question here is whether the authority for goal selection comes from a political/technologic elite group or whether it is determined through informed collaborative participation of those whose lives will be affected by whatever policies are adopted. It would be easier to write an objective stating that change agents will encourage citizen participation on a consumer council; however, such a statement describes behavior of a change agent rather than behavior of the client system. A more appropriate objective would be, "Citizen voting on consumer issues presented to Metropolis City Consumer Council will increase 20% within a three month period."

2. *Behavioral objectives should include an active verb that indicates behavior the client system will show in dealing with the proposed change.* The objective should not consist of a list of the proposed changes. For example, the stated objective, "decreased health services at the Central Health Office will be remedied by increasing financial support," does not describe the client system's action, only its needs. To put this idea into an objective, the following more adequately states actions involved: "The Central Health Office will expand its financial base through increasing its mill rate allotment from 0.5 to 0.8. Other objectives will discuss how the Central Health Office will act to increase its mill rate allotment, which complies with yet another characteristic.

3. *Objectives need to include only one behavior per objective.* Objectives containing more than one process element are difficult to deal with, as one part of the objective may be reached and the other may not be reached. For example, in the previous objective, inclusion of action statements about meeting with the financial board and presenting supportive data to increase a mill rate allocation would have made a long unwielding objective. Each of the processes is different (gathering supportive data, presenting data to financial board, and increasing a mill rate by 0.3 point) and needs to be stated in separate objectives.

4. *Objectives need to be ordered sequentially.* Although alternatives are a must for each objective, and logical order has a way of disappearing in complicated change strategies, it is necessary to list objectives in such a manner that movement toward the general goal is seen through each objective. Essentially, behavioral objectives are step-by-step directions for reaching a goal.

5. *Behavioral objectives should be realistic in terms of time available and characteristics of the client system.* Note that the example used in discussing the first characteristic included a terminal time for accomplishment of the objective.

Defining objectives has lead us to the second element in the planning phase, that of solutions.

Solutions

There are two components that comprise the element of solution: delineation of possible alternatives and the selection of appropriate actions.

The system term *equifinality* is especially meaningful during this phase of planning. There is rarely one avenue for reaching an objective. Alternatives are the various ways that the available resources may be allocated to achieve the desired results. There are multiple means of delineating the possible alternatives. One method that is often helpful is "brainstorming." The team can list every possible alternative, perhaps on a blackboard where they can be seen by all team members and openly discussed. Discovering what others have done is useful. This is not to say that what another agency has done would be transferrable, but why reinvent the wheel? Each alternate route needs to be examined and weighed in order to select the most appropriate action for the identified situation.

The selection of the appropriate action has opened up an entire new area known as operations research. *Operations research* is a broad term for a variety of methods developed to study a system and experiment with alternative solutions to problems that will lead to achievement of the desired objectives. Some of these methods are queuing theory, simulation, linear programming, and network analysis (PERT).

Queuing theory is a mathematical computation that enables the system to predict how the system will react to the demands placed upon it. This method is especially useful in situations in which there is a demand for service, for example, an outpatient clinic. There are three factors for consideration, the facility, the time of client arrival, and the time required to satisfy the client. For the outpatient clinic, data need to be collected concerning the average arrival time (whether everyone comes at once, or arrivals are staggered), the average serving speed, and the queue discipline [how are they served, first-in-first out (FIFO), random, or perhaps last-in-first-out (LIFO)]. Important also is the number of servers; this determines the number of queues. By means of mathematical formulas, the optimum number of clients that can be served in X amount of time can be computed, or the client time spent in the system, the length of time devoted to waiting, or the number of clients waiting. Theoretically, the arrivals should be less than capacity; otherwise, as the arrivals increase over capacity, the system becomes unmanagable.

Simulation is another means by which alternative solutions can be tested. This is the creation of systems under given conditions. The detailed characteristics of the system and the relevant variables are fed into a computer programmed to give the answers to what happens "if." If it is assumed that the quality of the data detailing the model is accurate the answers to the questions asked can then be applied to a larger real-life system. Advantages to simulation are obvious, the time element for revealing the consequences of actions as well as "injury-free" or "expense-free" testing. The use of computer simulation is relatively painless to both the client and the staff, and the method can reduce costly errors.

Linear programming is another mathematical method used to discover the best way to allocate resources while minimizing cost.

Network analysis (PERT) is a tool that can be applied to any area that requires a coordinated effort within a time-limited framework. Program Evaluation Review Technique (PERT) is an example of such a tool. The utilization of PERT assists the planner in making the decisions required to complete the objectives on time. Essentially, PERT divides the program into events (significant milestones) and activities (tasks related to events). The network then displays the relationships between events and activities. The network should be sequential for events to take place in logical order. The activities reveal the time required for completion and the required resources. An examination of a PERT network will reveal which activities have an excess of resources and the active possible trouble spots for tradeoff for main-tenance of the completion date. PERT has the advantage of visualizing the total program with events and activities.

The methods briefly described above are all examples of mathematical methods of selecting alternatives. Another means of testing an alternative is through a *pilot or demonstration project*. These projects provide the realism of a larger system but on a smaller scale. Obviously, they can only be implemented if the resources and time are not constraining factors. Model ward units or model health stations are some examples whereby changes can be implemented and evaluated and the results extrapolated to the larger system. Any of the methods used to evaluate or test possible alternatives are only as useful as the accuracy of the data used in the analaysis.

IMPLEMENTATION

The planning has been done and now the action begins. How do we implement? First, we are assuming (in this section) that the assessors, the planners, and the implementors though not necessarily the same person are interrelated and have a cooperative collegial relationship. Planning is worthless if implementation is not carried out in a manner that will achieve the desired results.

Once goals have been set, objectives stated, and alternatives outlined, the stage has been set for action. Action, however, does not automatically follow. Professional journals have been full of studies which assessed needs and presented realistic ways to meet these needs, but those in the position to follow through with action have done nothing. Behavioral scientists have been seeking answers to questions about human motivation. What processes help people to change as well as what processes operate to keep change from occurring? Bennis[1] described seven styles of change which have been used to bring about action. These are indoctrination change, coercive change, technocratic change, interactional change, socialization change, emulative change, and natural change.

Indoctrination change occurs where there is mutual and deliberate goal-setting under a one-sided power structure. *Coercive change* involves one-sided goal setting with deliberate intentions of using one-sided power. Chinese "brainwashing" and thought-control measures are examples. In *technocratic change*, one party defines the goal and the other party helps reach the goal without questioning the goal's value. Health professionals often expect this type of change to occur when dealing with clients. *Interactional change* has a shared power but occurs under conditions in which goals are not deliberately sought. The raising of children under the influence of parents who unilaterally define the goals is an example of *socialization change*. Here, there is one-sided power but collaborative goal implementation. *Emulative change* involves unilateral power without deliberative goals. Formal organizations where followers seek to act like their leaders is an illustration. *Natural change* is a category containing these changes which occur as a result of accidents, unexpected events, etc. It is characterized by shared power with nondeliberative goal-setting. Bennis suggests through this typology that change can be initiated by using various power distributions. There is an eighth category of change approaches that is known as Planned Change. This is the approach of choice for health professionals as it involves shared goals of the change agent and the client system. It also advocates shared power. Some would agree that the professional always has the power edge because of role status implications. The recent consumer's movement, however, has done much to equalize the power source. Comprehensive health planners probably had this in mind when advocating that management boards have at least 51 percent consumer participation.

The process of planned change involves a change agent system and a client system would work together applying valid knowledge toward solving a problem. Lewin[4] conceptualized the process of change as having three steps: (1) unfreezing, (2) moving toward the new level, and (3) refreezing at the new level. These three steps can be looked at in relation to the systems model presented in this text. Unfreezing involves the opening of the feedback loop to new variables of influence and/or to new patterns of data flow. Moving toward the new level involves interaction of the new data with the components potentially to change the system in terms of its calibration, organization, and/or equilibrium. Refreezing occurs as the system differentiates what is now part of the system and establishes negentropy. There have been change strategies developed to facilitate movement through these steps. Some of these strategies will be discussed as they relate to the systems model.

The straight line model presented in Figure 6 of Chapter 1 more clearly identifies points at which the change agent needs to make use of change strategies. The change agent must first recognize that resistance to change is always operational in any situation where the status quo is being shifted. The model depicts this as Environmental Input Exchange (Fig. 2, step 11, Chap. 1) which interacts with an internal systems structure, channels (2). As resistance is a normal occurrence in any system, the change agent must be able to categorize the resistance experienced into levels. Extremely high levels of resistance should be handled very gingerly. Lower resistance levels should be approached in such a way as not to increase manifestations of resistance. Resistance levels in social systems can be identified through behavior. Resistance behaviors of a social system increase proportionately to the degree the system perceives the change as threatening. Resistance levels increase when direct pressure for change is exerted. Although temporary changes can be brought about through use of direct pressure, the changes become surrounded by increased tension states. With such an unstable situation, major changes can occur suddenly and often unpredictably. Should the change agent wish to use such an approach, the dangers should be weighed heavily. The changes which finally occur may not be the planned ones.

Resistance levels decrease when the client system perceives the change as having the support of others who are trusted. Resistance levels decrease when the client system can "see how a new equilibrium is better than the old." Resistance also decreases when the client system has the opportunity to participate in the decision to change and in the implementation method used for change. With these considerations in mind, the change agent could make use of the following strategies. Select an interaction pattern which allows for system mutuality. *Synchrony* (14) or the bolt of lightening stimuli pattern does not allow the client system much opportunity to share most fully in the *interdependent interaction* (15) process. *Reciprocy* (14) and *helicy* (14) are the two patterns allowing for the most interaction of the client system. *Resonancy* (14) is a stimulus pattern to use once a client system has responded with a low resistance level to a pull of incoming change data. See Figure 6 of Chapter 1 for steps 14 and 15. Once the client system accommodates a wave length of data, use it to work *reciprocy* and *helicy*. Use of the least threatening most trusted component of the client system to identify and assess needs and to present planned change proposals requires strategies to decrease resistance.

Although change can be directed toward any level of internal systems functions, it can be seen that change directed mainly at altering the systems purpose will get little response from the client system. The purpose and negentropy (the most steady state of the system) occur together. Change is better directed toward lower and less stable internal systems processes. Seeking to increase the degree of nonsummativity among components is a more realistic point of change intervention. Systems in crisis are experiencing entropy. Intervention to facilitate new roles, normalcy, rules through collaboration, organization and/or equilibrium changes is another sound intervention point.

Return to the situation presented in Chapter 3. A nurse,[6] as the first Community Organizer for Mental Health in Israel, was presented with several community systems problems which could benefit from change. The following account outlines her approach to one of the problems—the building of an activities center.

The mayor was thinking, "This building will stand for my son's memory. It must be the best money can buy. It will have meaning to those living here and they too will have a piece of my son—like if he were somehow here after all." With this purpose in mind, to establish a memorial to his son, he proceeded to hire architects, builders, and contractors. None of these workers were of the village members who would go to the activity center. No one thought to discuss the plans with these Jews who came from some far away Eastern country.

The people were thinking, "This building has no piece of us in it. Hammer, hammer, noise all day and we sit with nothing to do." The acting-out boys did not think. They acted. Each morning some piece of the previous day's work was found undone.

In time the mayor was thinking, "Who are these bad Jews who tear at my son?" and the people were thinking, "No one in this place cares about us. We should be home in Iraq."

Such thoughts could not go on if these people and this town were to grow together. Yet, I could not go to the mayor and say, "All you have done in memory of your son is useless," nor could I go to the people and say, "Your children are acting-out because of your own unhappiness being without your families left in Iraq."

What I did was have a women's meeting. The woman from the village helped talk with the other women, encouraging them to come. Native foods were brought to be shared. Staff from the Health Station were sent invitations to come sample the excellent food. The mayor's wife was also invited. It was like a holiday. When the mayor's wife came the lady from the village, the nurse from the Health Station and I walked and talked together. We spoke of dreams and shared hopes about ourselves and our families. I asked, "What do you think about the new building? Will you (the village women) go there with your family and will we see you (the mayor's wife) there? Each looked at the other and the village woman said, "No, my family will not go there. We will not be wanted!" The mayor's wife learned that the village people felt left out of the project and that they thought the way it was being built would not include places for activities their families enjoyed.

Two days later all the noise at the building stopped and the workmen drove away with their tools. Soon the mayor was seen in the streets in the evening walking and talking with families. There was a meeting. An architect came to the meeting. The building plans were changed. When work started again some of the older boys worked with the builders. On weekends and some evenings the men helped clean the building site. The women sometimes brought cool drinks to the workers. The night time undoing stopped. When the building was done it was in constant use. [6]

The change agent in this situation made use of several change strategies. She did not attempt to change the mayor's purpose for the building nor did she confront the village women at their weakest point—loneliness and homesickness for Iraq, and concern over their acting-out children. She worked instead to increase the degree of interrelatedness or nonsummativity among the community system components. Her intervention began in a nonthreatening environment. The person who presented the needed data for change to the mayor was his own wife, apparently seen as a "trusted other." The major pattern of stimuli used was *reciprocy* which then developed into *helicy* with the continued meetings.

Other change agents may approach such a situation differently but the results could be equally as good. The concept of equifinality is always operational for change agents, and although what is done is done from a planned change model, results cannot always be predicted. This chapter has attempted to provide tools which can help the change agent plan and intervene most effectively. Even with such tools, change remains a risk-taking operation. It is our hope that nurses and other health care disciplines will be willing to take this risk in order to improve health care services offered to all.

REFERENCES

1. Bennis W, Benne K, Chin R (eds.): The Planning of Change, 2nd ed. New York, Holt, 1969
2. Bergwall D, Reeves P, Woodside N: Introduction to Health Planning. Washington DC, Information Resources Press, 1974
3. Hanlon, J: Public Health: Administration and Practice, 6th ed. St. Louis, Mosby, 1974, pp 243, 244

4. Lewin K: Quasi-stationary social equilibria in the problem of permanent change. In Bennis W, Benne K, Chin R (eds.): The Planning of Change, 2nd ed. New York, Holt, 1969
5. Palmiere D: Types of planning in the health care systems. Am J Public Health 62:1112-15, 1972
6. Riss B: Personal communication, 1975

RECOMMENDED READING

Arnold M: Basic concepts and crucial issues in health planning. Am J Public Health 59:1686-697, 1969

Bachrach L: Developing objectives in community mental health planning. Am J Public Health 64:1162-63, 1974

Baum M, Bergwall D, Reeves P: Planning health care delivery systems. Am J Public Health 65:272-75, 1975

Colt A: Public policy and planning criteria in public health. Am J Public Health 59:1678-85, 1969

Dror Y: Ventures in Policy Science. New York, American Elsevier, 1971

Federal Electric Company: A Programmed Instruction to PERT. New York Wiley, 1963

Flagle C: The role of simulation in the health services. Am J Public Health 60:2386-94, 1970

Fonaroff A: Identifying and developing health services in a new town. Am J Public Health 60:821-28, 1970

LeBreton P, Henning D: Planning Theory. Englewood Cliffs, N.J., Prentice-Hall, 1961

Makower M, Williamson E: Operational Research. London, English Universities Press, 1967

Merton W: PERT and planning for health programs. Pub Hlth Rpts 81:449-51 1966

Nathan M: Some steps in the process of program planning. Am J Public Health 46:68-73, 1956

Reisman A, Kiley M: Health Care Delivery Planning. New York, Gordon and Breach, 1973

Strauss M, deGroot I: A bookshelf on community planning for health. Am J Public Health 61:656-79, 1971

Williams J: Comprehensive health planning: an organizational means for transition. Am J Public Health 59:48-52, 1969

Zukin P: Planning a health component for an economic development program. Am J Public Health 61:1751-59, 1971

5
EVALUATION

The fourth phase of the decision-making process is that of evaluation. By definition, there appears to be little difference between evaluation and assessment. Each is considered a process by which some value is determined. Each has the word appraisal as a synonym. Frequently, the terms have been used interchangeably. Operationally, however, in reference to health care, we believe there is a difference. Assessment is used as a process to determine what is present or available. Evaluation may be thought of in terms of comparing or contrasting. Where assessment should be nonjudgmental, evaluation answers questions such as: Were we successful in meeting our goals? How well did we accomplish the task? What were the reasons for success or failure? Evaluation does indeed measure the value or worth of an action. Thus, in operation, evaluation (measurement of actions) differs from assessment (data-collection) even though the basic definitions are similar.

In an attempt to clarify and put into operation the term, the American Public Health Association[1] in its "Glossary of Administrative Terms in Public Health," gives the following definition:

The process of determining the value or amount of success in achieving a predetermined objective. It includes at least the following steps: Formulation of the objective, identification of the proper criteria to be used in measuring success, determination and explanation of the degree of success, recommendations for further program activity.

An examination of this definition provides us with guidelines for and

119

implications of the evaluation process. First, we are to measure the amount of success. This measurement could result in negative as well as positive information. For the future, knowing that an action is ineffective is as important as knowing that an action is effective. Knowledge of ineffective methods can save money, time, and effort for others, especially if the reasons are well-documented. Second, the measurement is to be of predetermined objectives. Inherent in this statement is the determination of how well the assessment and planning phases were accomplished.

In considering programs that affect community health, we have delineated three categories for examination. The elements or categories are interrelated and the total picture cannot be gained without an examination of all three. An administrator, here used as a term for the ultimate decision maker, needs the total picture to make appropriate programmatic changes. The categories of the program to be evaluated can be diagrammed as in Figure 1, which demonstrates the relationship between systems and places activities in proper perspective. Organization occurs in a system to meet a purpose. That purpose is translated into action directed in some way (to, for, with) another system resulting in impact on that second system. The action or activity occurs in the interface between the two systems. Subsequently, the impact may result in change in the originating system (organization), depending on the manner in which the system is structured to accept or receive imput. It is possible, then, that the impact may be encoded as deviation-correcting or deviation-amplifying data by the system. During the remainder of this chapter, each category will be discussed in more detail.

INITIATING SYSTEMS

Initiating systems have two program elements. These program elements are (1) organization and (2) resources.

Organization

In this context, organization is loosely defined as the "how" of accomplishing the task. This category contains four elements: philosophy and objectives, manpower, and resources. An examination of the philosophy and objectives of a program should be an initial step in program evaluation. A critical review will reveal the two elements are congruent. An example of noncongruence would be a philosophy that speaks of a commitment to community-based health care, but all the objectives are relevant to acute hospital-based care. The objectives should be clear, concise, and state what

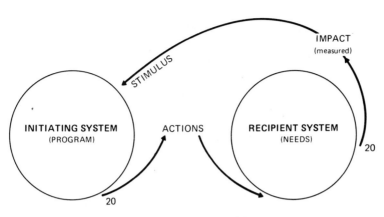

FIG. 1: Evaluation model.

levels of accomplishment will be satisfactory, as well as the time-frame for accomplishment of the task. Thus, a review of level of achievement and the time-frame will provide data as to priorities and if the objectives were realistic both in magnitude of expected outcomes and time for completion.

Manpower refers to the people involved in the program. Perhaps more important than the number of persons required are the functions and abilities required to accomplish the task. The ratio of number to function is vital. For example, you would not expect the manning document of a program serving 125,000 persons with specialized needs to show a majority of persons whose abilities enabled them to function at a level that could not possibly meet the program objectives. In addition to the manning document, a review of job descriptions can reveal a great deal as to how persons are expected to function. The usual case with job descriptions finds that their broad general terminology renders them a useless evaluative tool. The functional job description with expectations and standards included is being examined and accepted by many. A comparison of the functional job description with the actual functions of the individual or group reveals gaps between what the system perceives as the ideal and what is acceptable as normal rules of behavior.

Resources

Resources refer to physical as well as fiscal. "Physical resources" implies the building or where the program is based. Is it adequate and appropriate? Within the unit, is the physical plant structured for efficient client-flow? Has the facility been accredited by any national body as meeting some form of minimum standards? "Fiscal resources" are the budget. How much is available and how have the funds been allocated? Is the total allocation by line-item or has the total been subdivided by specialty areas (adminstration, medicine, nursing, etc.) or by specific objectives. Cost accounting is one method of evaluating fiscal resources. Home health care agencies which apply for accreditation by the National League for Nursing have to complete a lengthy complex cost analysis. The end result, however, demonstrates cost per visit by category of staff and enables fees to be set more realistically.

ACTION

Actions are purposes and objectives translated into operational activity. The term implies movement and meaningful behavior. In the community, the actions occur in the interface between two or more systems. The complexity of the actions depends on the purposes, objectives, etc., of the initiating system. These actions provide a stimulus to the target system. Measurement of actions includes what was done (quantity) and how it was done (quality).

The measurement of quantity is not foreign to those of us who have been associated with health agencies. Time and Effort Reports have been required for years. In academia, a compilation of faculty workload data is an example of quantitative measurement. Numerical computations of how many visits, how many clients seen, how many injections given, or how many credit hours taught are relatively easy to obtain. Many professionals, however, are not cognizant of the full impact of these computations. An often heard complaint is that such recordkeeping is unnecessary and time-consuming. Professionals rationalize the time required for the keeping of records could be better utilized providing services. Quantity of service (data collected from records), however, is one method utilized for justifying our professional existence. In this age of budget allocation, it may well be the group that can justify its existence as a relevant system component that will survive.

Measurement of quality of actions is much more complex. Measurement of quality of service infers setting of values and the collection of subjective data. What is a quality action, what are the standards or criteria by which an action will be measured, and what methods of measurement are valid? It is because of the difficulties in answering these questions that quantitative measurement has been used more frequently than qualitative measurement in evaluation. Another difficulty is the nontransference of evaluation data. Few programs exist across the country that have interrelated evaluation processes. One of these programs, Minnesota Systems Project, University of Minnesota, under the leadership of Dr. Vernon Weckworth, was designed as the evaluation arm for the 69 Comprehensive Children and Youth Projects. Even though the projects were individual and allowed to develop their own unique characteristics, data were collected that were common for all projects. Thus, data have been of assistance to administrators planning similar projects ten years later. However useful the data were, they were still quantitative rather than qualitative. It remained for the individual projects to measure their actions qualitatively. In multidisciplinary programs, it is the responsibility of each professional to design the criteria by which they will be evaluated.

Nursing has developed some tools for the qualitative measurement of nursing care. The one most frequently utilized is the nursing audit. Fundamentally, the nursing audit is a case-record review accomplished by peer review. The method can be utilized in a variety of health care settings. Other professionals use the audit system. As yet, however, there has not been an integrated audit tool developed. Perhaps the future will bring an audit tool that measures the quality of health services compared to the needs identified. For individuals, the Problem-Oriented Record holds promise. The utilization is still limited with regard to families and communities. The University of Michigan and the University of Texas, Schools of Public Health are developing and testing records for these areas of concern.

RECIPIENT SYSTEM

The recipient system response to these "act-ons" is through acceptance or resistance to the implemented change or innovation. Without some knowledge of these concepts, evaluation of impact leaves the investigator with the statement, "it worked (or failed), but I don't know why." Without this knowledge, adaptations cannot be made that will lead to program improvement. Goodwin Watson[3] categorized the factors related to resistance to change under the three categories of (1) who brought about the change, (2) what type or kind of change was initiated, and (3) how was the change implemented.

Who Brought About the Change?

Research on the acceptance of certain pharmaceutical agents revealed that the rate of acceptance was dependent upon the endorsement of competent authorities. Organization resistance to change is often less if consideration is given to involvement of lower echelons within the hierarchy in the planning and if the power structure has given support. In larger social systems, like a community, was the change the result of the imposition of power (legislation) or of individual voluntary associaton?

What Kind of Change Was Initiated?

Research has shown that acceptance of change has been greater and conversely resistance less if (1) the change could be seen as having positive attributes and (2) the change was not in direct conflict with the accepted cultural values of the recipient system. Positive attributes were also affected by factors such as increased costs or increased burdens. Were there positive results that could be seen and what were the visible payoffs? Change that is congruent (compatible in form, use, function) with the existing culture will be more readily acceptable. Anything else will be looked at with suspicion, viewed as a threat, and will never achieve the degree of internalization necessary for continued growth and development. Another contributing factor is the amount of change required for acceptance. Is the change one that can be adopted in stages or is it a change that will exemplify the statement, "new brooms sweep clean." The greater the magnitude of change, the greater the stress and anxiety induced, and the more the individual or group may feel threatened. Research has shown that modifications of the familar are more easily accepted than changes that represent radical innovations.

How Was the Change Implemented?

Acceptance or adoption of change has been shown to be greater if the decision for change was made through consensus rather than direct coercion. The importance of involvement of many in the total process cannot be overemphasized. Mechanisms should be built in to allow for feedback from the recipients.

IMPACT

Impact is the outcome of the action upon the recipient system. Measurement of impact reveals the degree and response of the system to the stimulus (action). There are two elements to be considered in measuring impact: change and satisfaction. Again, as with the other categories, impact measurement presupposes effective, appropriate assessment and planning. As qualitative measurement requires standards, impact (outcome) measurement requires baseline data to measure change or satisfaction.

Objective means for measuring impact (change) are relatively easy to obtain. Numerical computations of before-after data such as morbidity/ mortality data have become standardized. Therefore, state and county, as well as international, data can be compared, contrasted, and predictions made for the future. Another objective method is utilization review. This involves a numerical computation of how the service or product is actually utilized by the consumer, the assumption being that an increase in utilization patterns infers a positive change.

Until recently, the objective domain of measurement received the greatest attention. This was the area that was standardized. With the emergence of new health care delivery systems and the inclusion of the consumer as a partner in planning, the subjective domain (satisfaction) was included as part of evaluation. The measurement of impact (satisfaction) is more complex and leads the investigator to collect and analyze subjective data. As comprehensive health care centers began, we saw the increased emphasis on evaluation of client satisfaction. An outgrowth of the comprehensive health care center with its multidisciplinary approach was the emergence of new roles for health care providers. These providers were eager to demonstrate their effectiveness in their new role or in their new setting. Among the providers included were nurses with expanded skills such as the Pediatric Nurse Practitioner and the Family Nurse Practitioner, the Home Economist, and paraprofessionals such as the Home Health Aide. The difficulty with the subsequent data is the noncomparability. We look to the future again for standardized measures of client satisfaction that will make measures of impact comparable, capable of being contrasted, and predictable.

RESULTS OF EVALUATION

The result of impact measurement upon the initiating system, whether change or satisfaction, is dependent upon the level at which the initiating system will accept feedback. Many systems refuse to accept feedback that does not fit their expectations. This results in the following statement, "It's not the program, the client won't accept anything we do for him." It is this type of response that led Ryan[2] to write *Blaming the Victim*. Refusal to alter functioning as a result of feedback from the recipient system is an example of a deviation-counteracting-mutual-causal relationship.

Accountability

Another result of the evaluation of health care services is that of increasing accountability. The evaluation process is not complete unless the results are appropriately communicated. As members of the health care professions, we speak to the concept of professional accountability but what do we mean? What is accountability and what problems are related to accountability in community health?

Accountable, the adjective form of account, has been defined as "being answerable." Relevant synonyms are responsible, liable, (to) justify or explain. The health care system, by virtue of this definition, is responsible and liable for, and needs to justify and explain actions taken in the delivery of care. The definition implies several types of accountability. Responsible actions might be considered as those actions which were appropriate, effective, and efficient to meet the stated needs. Broadly interpreted, this area includes quantity of service and fiscal responsibility. Liability for actions infers safe practice. Justification and explanation of actions implies communication between the concerned parties.

The definition as stated and applied holds with the premise that we do not function in a vacuum. The health care system is answerable. But, to whom? Traditionally, the individual health care provider was accountable to self and peers (individual conscience and practice acts) and to organizations if employed (periodic evaluation). The organizations were accountable (fiscally) to funding agencies and/or governing bodies. But to "others" the commonly held belief was that either the "others" were not interested, concerned, or would not understand. As the consumer, as a group, began making an impact on planning and action, the cry went out for increased accountability from the provider, whether individual or group. The consumer was no longer satisfied with platitudes or professional mumbo-jumbo. The consumer wants to know what he is getting for his money. The consumer is interested and concerned about costs, quantity of services, as well as quality of care. Table 1 gives an example of accountability in public health from the vantage point of a program administrator. This table demonstrates the broad variety of persons, areas, and purposes underlying accountability.

Why should we be accountable? Why is accountability becoming increasingly important? As individual providers, we should welcome the opportunity to evaluate our services and to justify our existence as system components to others. Accountability for safe practice assists in deleting the unsafe practitioner from the professional ranks as well as raising a barrier to deter the admission of unsafe practitioners to the professional ranks. Safe practice is an element of quality care.

As components of the system, we have a responsibility to ourselves and to the community to communicate goals and discover gaps in service. Some components, such as official state agencies, are accountable because of legal responsibilities. An all-encompassing purpose for accountability is its justification for funding, especially in these times of tight monies.

The reasons for the apparent lack of accountability of the health care delivery system have been varied. Essentially the reasons can be categorized into three broad areas: lack of resources, lack of defined responsibility, and other reasons. Lack of resources as a barrier to accountability includes lack of specific measurable objectives and specific standards or criteria. Interpretation is difficult or impossible if there is no existing measuring unit. This barrier is directly related to educational experiences. Until recently, educational experiences were not structured to instill within the learner a philosophy that included the broad definition of accountability. Nor was the student given the opportunity to utilize the tools or methods required to demonstrate accountability.

Concomitant with lack of resources has been the lack of defined responsibility. No one was willing to say, "The buck stops here." Some of the barriers in this area were related to divided loyalties and vested interests, either individual or professional. Would you freely communicate data that might be ultimately damaging to either your profession or career? A primary barrier in this area is resistance to change. Adaptation to change is difficult for many. The process is always anxiety producing. Job security and status are often threatened. If a result of being an accountable professional meant disbanding your agency, could you cope with the situation or would you ignore the issue and hide the facts and proceed with business as usual?

The other reasons are linked to these two types and are alibis for apathy. The usual ones most often mentioned are: lack of money, lack of staff, or "no one has ever required this before," and "let's keep the status quo." These alibis inspired an anonymous author to the following:

Seven Steps to Stagnation

1. *We're not ready for that*
2. *We've never done it that way*
3. *We're doing all right without it*
4. *We tried that once before*
5. *It costs too much*
6. *That's not our responsibility*
7. *It just won't work.*

Luckily, the alibis are being scrutinized and are no longer considered adequate "reasons" for inactivity.

The future is going to bring increasing emphasis on accountability. The realization of facing unlimited needs with limited resources is facing everyone concerned with community health. Therefore, we need to define the problem clearly; plan for effective, efficient, quality services that include effective utilization of personnel; coordinate activities to avoid unnecessary duplication of services; demonstrate programmatic impact or change in health status; and above all use language that can be understood by those with whom we are having discourse and dialogue.

TABLE 1. ACCOUNTABILITY IN PUBLIC HEALTH.*

Responsible Party	Accountable to Whom	In the Area of	For the Purpose of	Measurable by
Program administrators	Public at large	Communicable disease Prevention	Attaining acceptable standards of quality of life	Mortality, morbidity
	Individual consumer	Services provided	Quality, quantity, adequacy of services provided	Mortality morbidity
	Providers (participants)	Fees, roles in decision making, planning, working conditions	Acceptable life-style, prestige, self-esteem	Provider feedback staff turnover
	Governmental bodies	Effectiveness, efficiency, adequacy of program	Satisfaction of constituency; funding, legislation	Fiscal, mortality, morbidity assessment; favorable public image
	Parent organization	Program priorities, efficiency, effectiveness	Organizational and financial support to meet consumer needs	Mortality, morbidity, administrative guidelines, standards

Sister organizations	Communication, cooperation, program assistance	Coordination of efforts to meet community needs; stimulate community interest; better public image	Formation, frequency of meeting of joint councils; degree of cooperation in joint projects
Professional organizations	Service priorities, patient eligibility, service components, support of professional organizations	Professional recognition; professional and political support	Professional organizations endorsements and cooperation
News media	Communication; cooperation	Maintaining public image; political and financial support	Media's support of agency's efforts
Special interest	Cooperation with efforts in their special interests	Public image, financial and political support	Degree of cooperation sought by special groups

*Adapted from: material presented at 41st Annual Meeting, Southern Branch American Public Health Association, Louisville, Kentucky, May 9-11, 1973. Topic: Your Accountability for Health—How Much is Lots?

With the benefits to be derived from an objective evaluation, why is evaluation sometimes viewed with such hostility? There are a variety of reasons why evaluation is not utilized to the fullest extent. Economically, a program may give evaluation a low priority. This is especially true if the providers and administration feel the services provided are "essential." Other providers are so confident of their role as "quality providers" that any study that might undermine this confidence is in question. Another reason is the fear that evaluation might reveal serious lack of congruence between program goals and organizational goals. On the other hand, an evaluation might reveal a program has done so well that it is no longer needed. Even though evaluation was planned, the apparent success of a program often causes evaluation to be forgotten.

We have been considering the phases of the decision-making process as separate units, as blocks, which, when placed one upon the other, take shape, forming the whole. In reality, the phases are like interwoven strands that form a cable leading to a transformer. The transformer fails to function if there is a defect or a break in any of the strands. Thus, the decision-making process, to be truly effective, requires all four phases. Evaluation and action based on the refining and redefining of objectives are the means by which programs can be continuously adapted to meet the changing needs of society. Evaluation, reassessment, and reprogramming to meet current and projected needs give the system flexibility, continuity, and relevance. It is the process by which entropy can be avoided.

REFERENCES

1.   Glossary of administrative terms in public health. Am J Public Health 50:225-26, 1960
2.   Ryan W: Blaming the Victim. New York, Vintage (Random House), 1971
3.   Watson G: Resistance to change. In Bennis W, Benne K, Chin R: The Planning of Change, 2nd ed. New York, Holt, 1969, pp 488-98

RECOMMENDED READING

Adams A: How to Study the Nursing Service of an Outpatient Department. Washington, D.C., U.S. Dept. of Health, Education and Welfare. Public Health Service Pub. #497. 1964

Bernie R, Readey H: Problem-Oriented Medical Record Implementation. St. Louis, Mosby, 1974

Caputo D: Evaluating urban public policy: a developmental model and some reservations. Pub Adm Rev 33:113-19, 1973

Cost Analysis for Public Health Nursing Services. Methods I and II. New York, National League for Nursing. Dept. of Public Health Nursing, 1964

Donabedian A: A Guide to Medical Care Administration, Vol. II: Medical Care Appraisal. New York, The American Public Health Association, 1969

Gortner S: Scientific accountability in nursing. Nurs Outlook 22:764-68, 1974

Graham S: Studies of behavior change to enhance public health. Am J Public Health 63: 327-34, 1973

Greenberg B: Goal setting and evaluation: some basic principles. Bull N Y Acad Med 44:125-30, 1968.

Helt E, Pelikan J: Quality, medical care's answer to Madison Avenue. Am J Public Health 65:284-290, 1975

Hurst J W, Walker H K: The Problem-Oriented System. New York, Medcom Books, 1972

Katz E: The social itinerary of technical change: two studies on the diffusion of innovation. In Bennis W, Benne K, Chin R (eds): The Planning of Change, 2nd ed. New York, Holt Rinehart & Winston, 1969, pp 230-55

Keeler J: The process of program evaluation. Nurs Outlook 20:316-19, 1972

Lewis E: Accountability: how, for what, and to whom? Nurs Outlook 20:315, 1972

Myers B: A Guide to Medical Care Administration, Vol. I: Concepts and Principles. New York, American Public Health Association, 1969

Notes and distributed material from: 41st Annual Meeting. Southern Branch American Public Health Association. Louisville, Kentucky, May 9-11, 1973. Topic: "Your Accountability for Health—How Much is Lots?"

Oglesby M, Carl M: The development and evaluation of health care system: a heuristic model. Nurs Research 23:334-41, 1974

Phaneuf M: The Nursing Audit: Profile for Excellence, 2nd ed. New York, Appelton, 1976

Saba V: Management Information Systems for Public Health-Community Health Agencies. New York, National League for Nursing, 1973

Starfield B: Health services research: a working model. New Eng J Med 289:132-36, 1974

Suchman E: Evaluative Research—Principles and Practice in Public Services and Social Action Programs. New York, Russell Sage Foundation, 1967

Weckworth V: Evaluation: Tool or Tyranny. Systems Development Project, University of Minnesota, 1971

Weiss C: Evaluation Research. Englewood Cliffs, N.J., Prentice-Hall, 1972

Section III

THE FUTURE

... if you do not think about the
future, you cannot have one.

JOHN GALSWORTHY
Swan Song

Health care is influenced and shaped by the larger societal system. The
form the delivery system takes is dependent upon the type of societal
structure, the attitudes and values toward health, and the processes by which
societies manage the services considered imperative for survival.

The current humanistic revolution is doing much to reinstate the
individual to a high level of worth. This trend can also be seen in the health
care delivery system, where each *specialty* area states that it seeks to treat the
"whole man." The concepts of autonomy and self-actualization promote
individualization, which then feeds into the consumerism movement. Societal
status functions are being aggressively challenged. The physician is losing his
"great white God" image. Malpractice insurance costs are sky-rocketing
because of the increased number of law suits against the medical practitioner.
Cultural systems are showing the effects of previous irresponsible manipulation for societal and individual needs. Our "space ship" Earth may already be
on a self-destruct mission that no new scientific knowledge can stop. Rapid
change has been added to "death and taxes" as inevitable life experiences.

133

How will our culture/society systems deal with these additional stresses and strains? How will health care delivery systems cope with change demands from both smaller subsystems (clients) and the larger suprasystem (culture/society)?

THE COMMUNITY

In the past 20 years, growth of urban areas has occurred at such a rate and in such a manner that a new descriptive term has been coined—*urban sprawl*. Moreover, this increased urban size does not indicate an improvement in the quality of life. The promised megalopolis of the future is in sight and the sight is not pleasing.

With growth of the metropolitan areas and expansion of suburbia, the downtown areas have deteriorated. Over the past 10 years cities have become aware of these problems and attempted to change the emerging profile. This "face-lift" has taken a variety of forms. Some cities restore existing downtown facilities through renovation of buildings, additon of potted plants, and brighter night lighting. Some older residential areas in the inner city are being reclaimed through homesteading. Other cities are using urban renewal to fund model cities programs which decentralize services for increased accessibility. Still other cities strip out the center city and make it a highly specialized recreational and/or business convention area. Sports arenas, trade centers, and elite hotels become symbols of the new downtown.

States such as Florida and Oregon have examined their resources and decided to limit expansion. Oregon has an active antipeople program and is even willing to sacrifice the tourism dollar to maintain limited population growth. Bumper stickers from the area read, "A nice place to visit, but you wouldn't want to live here." In other attempts to guide growth, new communities—Reston, Maryland, for example—have developed from the ground up as completely planned communities. Communities of the future are being experimentally designed by architects to find solutions for space and energy utilization needs. Arizona is actively engaged in promotion of solar energy research with an eye to the time when the desert may not be so empty.

HEALTH AND HEALTH CARE DELIVERY

A brief overview of the health care delivery system was presented in Chapter 2 of this book to show some of the interrelated variables currently impacting on health care. In addition to those variables discussed as current influences, are variables which have their roots in history. Essentially, it is the culture/society system that places value on characteristics inherent in health or illness. It is also the culture/society system that establishes roles related to these values. The progress of health care can be traced through identifying the cultural/society system's values about wellness and illness and through describing characteristics of associated roles. Elements of these historical variables discussed earlier continue to exist in the health care system. Because they still exist they continue to have impact. Note the reemergence of psychic healing, an ancient treatment method, as patients seek treatments which require that they be seen and considered as individuals rather than as the "GYN in Room 1203W." Consideration of past and present trends can help provide a framework that will assist in predicting the future.

Fundamental to the issue of health and health care delivery system is the debate of whether health care is a right or a privilege. Rights stem either from established custom or society or from formalized laws. The preamble to the Declaration of Independence in its listing of inalienable rights included "life" as well as the "pursuit of happiness." A broadly defined concept of health, which envisions health as man reaching for self-actualization, includes more than absence of disease, and could be subsumed as being part of these rights. Two hundred years have passed and the author is not available to define specifically his meaning. We cannot help but ask if we are not depriving man of his "right to life" when available technology is inaccessible because of limited access or financial barriers.

Robert Sade, in a 1972 debate,[7] considers health care neither as a right nor as a privilege but rather as a commodity that can be purchased in much the same way as buying groceries or clothes. A critical review of his article discloses an equating of health care with medical care. Hildegard Peplau,[6] in an affirmative response to the question, Is health care a right?, affirms that the recognition of health care as a right is much like the awareness that led to the Civil Rights Act of 1954. The consumer is becoming more aware of inequalities existing within the health care system, and is aggressively questioning providers and institutions about his health and health care.

The American Hospital Association has endorsed a "Patient Bill of Rights," but it is questionable how widely this information has been disseminated and more importantly implemented. George Annas,[1] Director of the Center for Law and Health Sciences, Boston University Law School, has written, *The Rights of Hospital Patients: The Basic ACLU Guide to a Hospital Patient's Rights*, to assist the patient in his self-advocacy. Some institutions have created slots for the Patient-advocate or the ombudsman, but these positions are not common to all institutions. What concerns us is whether these individuals can truly be patient/client advocates when they are dependent upon the institutional hierarchy for their own professional survival.

The report of the President's Commission on Human Rights outlined the following four goals for achievement of rights to health as embodied in the Partnership for Health Act of 1966.[9]

1. *Every person should have maximum protection against diseases that need not happen and against illness and injury resulting from the hazards of the modern environment.*
2. *Every person should have ready access to basic medical care—despite social, economic, geographic or other barriers—and should have the assurance of continuity of quality service through diagnosis, treatment, and rehabilitation.*
3. *Every person should also have maximum opportunity for developing his capabilities in an environment that is not merely safe but conducive to productive living.*
4. *All activities conducted in pursuit of health should be carried out with full attention to the dignity and integrity of the individual.* *

Walter McNerney,[5] in the Tenth Bronfman Lecture to the American Public Health Association, challenged those present to reorder priorities and broaden the definition of health to include more than treatment of disease. Obviously, awareness of health in its broadest sense is spreading among professional groups organized to deal with health issues.

We seem to be caught in the middle of what *is* and the vision of what *might be*. Essentially, the problem revolves around unmet needs and demands growing out of rising expectations. Unmet needs and expectation demands can be categorized under the broad headings of social problems: barriers to utilization of services and system problems; problems resulting from within, types of services, personnel distribution, etc. The goals as stated by the President's Commission are noteworthy, but who has assumed responsibility for implementation of programs that would realize these goals? We suggest that each component of the health care system has a share of the responsibility. These components have been discussed previously as: consumer, provider, and institution. Examine these components in light of their attributes and responsibilities for making these goals a reality.

Consumer. The consumer, as a human being, has the capacity to inquire, learn, seek self-actualization, and enjoy the freedom of movement through time and space. These capacities, combined with autonomy, give faculty to participate in the decision-making process. If man has the right to health and health care, as a consumer he has certain coexistent responsibilities. These responsibilities include becoming informed, questioning and sifting information, seeking to find the best answers, availing himself of services, and providing feedback to the other components of the system. The consumer can assist the provider and institution to plan for health services in a variety of ways. Two of the most important areas for consumer impact are: (1) the consumer providing input for the identification of health needs as he perceives the needs, and (2) open lines of communication between consumer-provider-institution. This latter would lay to rest the many misperceptions and misunderstandings and make possible the redesign of more realistic services; it would also make possible the implementation of priorities. Those consumers already aware of their rights and responsibilities will need help in finding the most appropriate channels for their feedback. Consumers not yet aware of either their rights and/or responsibilities will need help in learning how to be informed, how to question, and how to sift information. Predictions for the future may seem uncertain but it is certain that the role of the consumer will increase in importance and power. Preparation for this more active consumer role should begin now.

Provider. The provider, as a learned role, has the responsibility for making available the best information, presented in a manner that is understood by consumers. Withholding of information is a mechanism for passive display of power. Providers involved in such games do little to fulfill their own role responsibilities. In the future, among these responsibilities will likely be the promotion of health care as a right for all and the reordering of priorities actually to fulfill the requirements of this right. Providers must also assume responsibility to cut across specialization barriers to bring an interdisciplinary approach to health care that can truly meet the needs of the "whole man." The provider has the responsibility for the recognition of health care as a right and to reorder priorities to assist in achievement of associated goals. Such a recognition would bring about an awareness that solutions depend upon collaborative, interdisciplinary relationships. With such an awareness would come a restructuring of the educational system that would encourage early collaborative relationships, thus making the transition to the real world easier. The consumer would be accepted as a full partner in planning the health care delivery system. Recognition of the consumer as a competent component and partner would do much to reduce some of the barriers to health services.

Institution. The institution has been given the role of controlling much of the system. The institution, whether governmental agency or health facility, has the capacity for controlling financing, facilities, legislation, and the responsibility for protection of rights. Institutions not only provide service to consumers but are often centers of learning for the providers. Strategically placed as they are, institutions have the potential to promote meaningful changes. Since most institutions are organized along rigid bureaucratic lines, however, their ability to change within themselves has been limited. The key to the future of the role of institutions seems to lie in the need of institutions to find more open workable organizational patterns.

Different models for provision of health care have been developing over the years. Prepaid group practice is not a new concept but is only now gaining acceptance. First trials in prepaid group practice in the 1920s and early 1930s were met with dismay and harrassment. The Kaiser-Permanente program has had 30 years of experience and may be the model for many of the now emerging health maintenance organizations (HMOs). Garfield[3] has suggested a delivery system model that separates the sick from the well by establishing an entry level health-testing service. The client would then be referred to the component (health care, preventive maintenance, sick care) most relevant to his condition. Flow of clients and information would be controlled by a computer. Physicians' time would be conserved through the wide use of "paramedical" staff in the nonsick care components. Leininger's model[4] focuses primarily on the client and community and emphasizes an interdisciplinary approach to health care delivery. As in Garfield's model, the sick are separated from the well and a variety of services are offered the client. In Leininger's model, however, the system is client-centered and utilizes fully all available health personnel. The proposed model would permit multiple entry and exit routes and the provision of care by the most appropriate professional.

A recent development is the Health Planning and Resources Development Act of 1974 that replaces Hill-Burton, Regional Medical Programs, and Comprehensive Health Planning. This new act is designed to include local review and approval of federal funds in a geographically defined area as well as a regulatory plan to cut cost. At present (Nov. 1975), the act will be implemented by 1976. The Act will allow for consumer as well as provider input and is organized at four levels: national, regional, state, and local. At present, there are 202 areas that have been designated as Health Planning Regions. Each area will have a Health Services Agency. These Agencies will have as some of their goals: (1) to improve the quality of medical care within the area, (2) to perform long-range planning with regard to services, (3) to review and approve health funds in the area, (4) to advise state agencies on certificate of need, (5) to review existing facilities and approve funding requests for new construction. At the state level, the State Health Planning Agency must have a state-wide plan that will include the above goals in relation to health services and in addition will be the regulatory body for certificate of need. An outgrowth of the State Health Planning Agency is the Coordinating Council that is responsible for integrating local plans into the state plan. The national office of the Bureau for Health Services and Resources will provide technical assistance for planning methodology and the criteria and standards for the various review areas.

What the future role of institutions in health care will be depends on their ability to take the future seriously and to make the necessary internal adjustments that will expedite ability to change in response to new demands and needs. Whatever models are implemented, it is imperative that evaluation be an integral part of the model system.

WHERE DO WE GO FROM HERE?

For those of us who are in the health professions, the years ahead hold the promise of being exciting, challenging, and changing. The image of the exact form or structure that will result from change is fuzzy. What is clear is that each of the components identified as being a part of health care will have to expand its roles and increase interaction and interrelatedness with other components. Toffler's[8] recent book, *Learning for Tomorrow,* is devoted to discussion of the need to develop "future-focused role images." Several chapters present concrete approaches to expanding role images. It was noteworthy that systems theory is among these approaches.

For some, systems theory has been interpreted as antihumanistic or mechanistic. We believe that it is only through observing the whole that we can anticipate or define our future roles. Capacity for forecasting, evaluating, anticipating, and reassessing enables systems to maintain openness, to be dynamic and responsive to needs. Systems theory provides a framework for interdisciplinary participation and decision-making. A systems approach to health care could be helpful in the future when we

> will be called upon to face novel situations which find no parallel in our past, and that therefore the fixed person for the fixed duties, who, in older societies was such a godsend, in the future will be a public danger.[2]

We believe that the future holds promise in the areas of policy making, planning, and goal setting. There will be collaborative planning through which individuality and creativity will be nurtured within a framework able to provide comprehensive services. Policy making, at whatever level, will be flexible enough to allow for local differences, to enable local implementation based on local needs. The formulation of goals and objectives will be based on assessment including the preferences and attitudes of the consumer group. Such formulation would result in activities using all available tools and resources and in services that meet societal needs. Collaborative, conjoint efforts would go far in closing the gap between the reality of what is and the dream of what could be.

Future implies change. The concept of change includes the elements of decision-making. A refusal to be involved in decision-making is a denial of our responsibilities. Let us hope and work for a change that is indeed planned. Let us synthesize the knowledge we have in the achievement of high-level wellness for our societal systems.

REFERENCES

1. Annas G: The Rights of Hospital Patients: The Basic ACLU Guide to a Hospital Patient's Rights. New York, Avon, 1975
2. Evans LH, Arnstein GE: Automation and education: hypothesis for a partial answer. Today's Education, 1962, p 3
3. Garfield S: The delivery of medical care. Sci Amer 222:15-23, 1970
4. Leininger M: An open health care system model. Nurs Outlook 21:171-75, 1973
5. McNerney W: Health care reforms: the myths and realities. Am J Public Health 61:222-32, 1971
6. Peplau H: Is health care a right? Affirmative response. Image 7:4-10 1974
7. Sade R: Is health care a right? Negative response. Image 7:11-19, 1974
8. Toffler A(ed): Learning for Tomorrow: The Role of the Future in Education. New York, Vintage Books, 1974
9. U.S. Department of State: For Free Men in a Free World: A Survey of Human Rights in the United States. Washington DC, U.S. Gov't Ptg Off, Publication #8434, 1969, pp 166-68

RECOMMENDED READING

Breslow L: The urgency for social action for health. Am J Public Health 60:10-16, 1970

Brown A: Rights, responsibilities, logic. J Miss State Med Assoc 12:492-93, 1971

Dahinden J: Urban structures for the future. Intellectual Digest, 1972, pp 69-72

David S: Society's responsibility for health. New Engl J Med 283:766-68, 1970

Eldridge HW (ed): Taming Megalopolis Vols. I. and II. Garden City, Anchor Books, 1967

Jones S: Consumer's role in community planning. Indiana State Board of Health Bulletin, 1970, pp 14-16

Leiberman EJ (ed): Mental Health: The Public Health Challenge. Washington, American Public Health Assoc, 1975

Lifton R: The Struggle for Cultural Rebirth. Harpers, 246:84-88, 1973

Lorig K: Consumer-controlled nursing. Nurs Outlook 17:51-52, 1969

Lum D: The health maintenance organization delivery system. Am J Public Health 65:1192-1202, 1975

McClure W: National health insurance and HMOs. Nurs Outlook 21:44-48, 1973

McNeil H: How to become involved in community planning. Nurs Outlook 17:44-47, 1969

Somers A: Health Care in Transition: Directions for the Future. Chicago, Hospital Research and Educational Trust, 1971

Symposium: Health care: Rx for change. Saturday Review, 53:17, 32, 1970

Terris M: Crisis and change in America's health system. Am J Public Health 63:313-18, 1973

Toffler A (ed): Learning for Tomorrow: The Role of the Future in Education. New York, Vintage Books, 1974

U.S. DHEW: Towards a Comprehensive Health Policy for the 1970's, May 1971

EPILOGUE

During the preceding chapters we have attempted to show you, the reader, how to "take a wider view, look at the whole landscape, really see it, and describe what's going on here." We hope you have not been lulled into complacency, thinking that the models as presented are the only answer, or that we have covered all the material. Rather, we hope to have whetted your curiosity for exploring other theories, challenged your creativity for developing models for practical application, and expanded your insight in a way that will lead to exciting new adventures just over the horizon.

In the words of Jonathan Livingston Seagull,

Don't believe what your eyes are telling you. All they show is limitation. Look with your understanding, find out what you already know, and you'll see the way to fly. *

*From Jonathan Livingston Seagull by Richard Bach (Text: Copyright © 1970 by Richard D. Bach. Courtesy of the Macmillan Publishing Co.).

APPENDICES

APPENDICES

Appendix A

GLOSSARY

Boundary	The specification of what is inside and what is outside a system.
Calibration	A change in number or amount within a fixed range of variations that occur in a system.
Closed System	A system in which there is no input exchange with the environment; currently the term is being used to refer to social systems which are characteristically rigid, have poor utilization of feedback, and in which change and growth are inhibited.
Components	The major structural elements of a system.
Cybernetics	A science that studies feedback loops by placing value weights on each piece of data selected or rejected by a feedback mechanism.
Deviation-Amplifying Mutual-Causal Relationship	A feedback loop segment that acts to increase deviations found in internal systems operations.
Deviation-Counter-acting Mutual-Causal Relationship	A feedback loop segment that acts to correct internal system deviations.
Ecologic Variables	Forces affecting the system that arise from the environment; four most general classifications of these variables could be:

1. Mobility characteristics—spatial-geographic,

temperature, availability of essential resources (water, nourishment or chemical, and/or other basic elements).
2. Characteristics of the coexisting population—number, type, size.
3. Physical characteristics—presence or absence of damaging elements.
4. Potency characteristics—intensity of any of the above factors.

A more specific classification of these variables as they relate to social systems could be: (1) spatial, (2) sociologic, (3) physiologic, (4) psychologic, (5) economic, (6) political, (7) cultural.

Energy Transformation	The process that changes energy from an inert form to an activated form.
Entropy	The tendency of all system functions to move toward disorder; also explained in the Second Law of Thermodynamics.
Equifinality	Term referring to the increasing complexity of interrelationships found in system functioning: the same final state may be reached from different initial conditions and in different ways; different final states may be reached from the same initial conditions.
Equilibrium	Normalizing segment of system function.
Feedback	Data from the environment that is introduced into and affects the internal systems operations.
Helicy	A pattern of stimuli characterized by a cumulative effect of stimuli inputs occurring between system and environment along a spiraling longitudinal axis bound in time-space.
Hierarchy	Arrangement of systems in order of complexities; arrangement of component roles in order of rank; the "organizational structure" of a system.
Inert Energy	An available source of energy or an energy pool.
Input	The energy or information from the environment going into and affecting the system.
Interface	The space or region between the boundaries of systems composed of a medium which can transport information, energy, etc., from the output boundary of one system to the input boundary of another system.

Macrolevel System	A large complex system the components of which may be subsystems whose components are systems (in some general systems theory texts, called a suprasystem).
Mezzolevel System	A system that includes a number of subsystems or microsystems.
Microlevel System	A small system whose component element is straightforward and uncomplicated.
Mutual Causal Relationships	The relationship the feedback loop segment has with internal systems operations.
Negative Feedback	Data which when accepted by a system lead to maintenance of stability.
Negentropy	The condition which describes the final steady state systems reach after differentiating among which feedback to include in continued system operations and which feedback to present as systems output.
Nonsummativity	Term that describes the degree of interrelatedness or wholeness within internal system operations—the gestalt effect.
Open System	A system with constant exchange of energy and communication with the environment.
Organization	The hierarchy of roles.
Output	The energy or information going out into the environment.
Positive Feedback	Data which when accepted by a system lead to change.
Primary Regulating Processes	Terminology meaning the same as the internal management system.
Process Elements	Those elements of system functioning which are visible through action or behavior.
Reciprocy	A pattern of stimuli which involves mutual interaction between system and environment.
Resonancy	A pattern of wavelike stimuli.
Roles	Name for classes of observable component behaviors in response to organized classes of stimuli.
Secondary Regulatory Processes	Terminology meaning the same as the feedback loop.
Stimulus	The "kick" that begins all system operations.
Structure Elements	Those elements of system functioning which act or are acted upon.
Synchrony	A pattern of stimuli presented to and accepted by a system at a specified point in time-space.
System	A complex of components in mutual interactions.

Appendix B

CURSORY ASSESSMENT OF A SELECTED SYSTEM

Name or type of system being assessed: biologic, physical, social classification and then specific name of system

Example: Social System—Emergency Room Service at Community Hospital

Physical System—A kidney machine

Biologic System—A kidney

Level of system complexity with rationale for selection: micro, mezzo, macro

PRIMARY REGULATORY PROCESS: characteristics of stimulus required to activate energy, strength, origin, intensity, source

Usual mainfestations of resistance

Component characteristics: identification by name or title; Description of the interrelatedness of these components in terms of their degree of non-summativity; compare and contrast characteristics with statistical norms

Role Characteristics: list of the roles seen in their order of hierarchy (a diagram may be useful); describe role functions

Rules: list of the apparent rules, and of the hidden rules

Coherence: clarity of roles and rules to all components; areas of obscurity

Purpose: stated purpose of the system; unstated purpose of system, if any; consistency of stated system purpose with assessed system purpose

SECONDARY REGULATORY PROCESSES: feedback loop; ecologic variables of influence: which rejected as input, which accepted, which presented as output, which withheld

151

Major patterns of stimuli noted: description of pattern (helicy, reciprocy, resonancy, synchrony) in terms of actual situations

Speculated system points receptive to change: usual change mechanism (organization, calibration, equilibrium)

Boundary: Unclear, well-defined, open, closed; description of events helps to determine selection

Other remarks:

Appendix C *
HISTORICAL CHRONOLOGY OF MEDICAL CARE IN THE UNITED STATES

The following pages present an historical chronology of the evolution of medical care in the United States. The chronology is developed in four significant periods, as follows:

Growth and Development of a New Nation (1781-1865)
Industrialization, Urbanization, and Problems of Poverty (1865-1929)
A Decade of Drought, Darkness, and Depression (1930-1939)
War, Peace, and Prosperity (1940-1969)

In each period, the events are presented under headings of Private Action or Public Action, according to whether the event was primarily related to private, voluntary, or professional action or to Federal, state, or local governmental action. This division emphasizes that both private and public action are important and necessary in the medical activities at critical times in the nation's medical care history.

Events were selected for inclusion in the chronology on the basis of their consequences for medical care generally, their importance as precedents, or their significance in reflecting trends, concepts, or issues. The chronology is therefore not all-inclusive but is intended to provide a broad historical perspective of medical care in the United States.

*From Myers: A Guide to Medical Care Administration, Vol. I. Concepts and Principles, 1969. Courtesy of American Public Health Association.

TABLE 1. EVOLUTION AND REVOLUTION IN THE COLONIAL PERIOD
(1620-1780)

Of necessity, the early settlers of our nation were confronted with the monumental task of individual survival in a hostile land. It is noteworthy that both private and public action in the medical care field was recognized as an important prerequisite to the achievement of success.

PRIVATE ACTION

1657: The Scotts Charitable Society of Boston was organized to help the sick, aid the poor, and provide burial for deceased countrymen.

CIRCA 1750: The Society of Friends (Quakers) organized several clinics for the provision of ambulatory care.

1756: The Pennsylvania Hospital (Philadelphia, Penna.) was the first institution constructed in the United States exclusively for the care of the sick. Voluntary subscriptions were responsible for its construction and continued maintenance.

1765: At the College of Philadelphia a professorship in the theory and practice of medicine was created which marked the beginnings of the first medical school in the United States. The school soon became affiliated with the Pennsylvania Hospital.

PUBLIC ACTION

1662: The first almshouse in the colonies for the care of the indigent was established in the Massachusetts Bay Colony.

CIRCA 1730: Almshouses were built in the larger colonial communities for which a physician was customarily engaged for a stipulated annual salary.

TABLE 2. GROWTH AND DEVELOPMENT OF A NEW NATION
(1781-1865)

From 1781 to the Civil War, a struggling new nation emerged as a power to be recognized. It is not surprising that important private and public action was taken in the organization of health services for population groups which were of prime importance to the developing nation (eg, merchant seamen), and later to correct harsh practices and attitudes, inherited from Europe, concerning the sick (eg, mental illness).

PRIVATE ACTION

1786: The Philadelphia Dispensary opened its doors as the first outpatient clinic in the United States.

1791: The New York Dispensary was founded and included outpatient services as well as home care.

1796: The founding of the Boston Dispensary, Boston, Massachusetts, was the first organized medical care service in New England, to insure that "the sick, without families, may be attended and relieved in their own houses . . . at a less expense . . . than in a hospital." This was the recognized forerunner of present day home care programs.

1840: Baltimore College of Dental Surgery was chartered and was the first formally organized dental school in the world.

1841: Dorothea Dix began a crusade for improvement in the treatment and care of the mentally ill. This resulted in the establishment of state hospitals for the mentally ill.

1847: The American Medical Association (AMA) was founded and began the formation of state and local societies. State medical societies immediately began appealing to their respective legislatures for registration laws. Medical education standards were among the first of the AMA's objectives.

1859: The American Dental Association was established as the first national organization of local dental societies.

PUBLIC ACTION

1793: The first local health department with a permanent board of health was formed in Baltimore, Maryland.

1797: In New York, the first Medical Practice Act was passed to regulate "The Practice of Physics and Surgery." This was the first act expressly to prohibit the corporate practice of medicine in the newly formed United States.

1798: The Marine Hospital Service was established by an act of Congress, to provide for the temporary relief and maintenance of sick and disabled seamen. This was the first prepaid medical care program in the United States, financed through compulsory employer tax and federally administered. This service later became the Public Health and Marine Hospital Service.

1841: The first law regulating the practice of dentistry in the United States was passed by the Alabama legislature.

TABLE 3. INDUSTRIALIZATION, URBANIZATION, AND PROBLEMS OF POVERTY
(1865-1929)

The revolution in industry, which took place in the middle of the nineteenth century in Europe, presented new opportunities to the nation as it moved ahead following the ravages of civil war. The consequent urbanization gave greater visibility to the health needs of the populace, and the related problems of poverty. Through private action and public support, the United States gave increasing recognition to the importance of a healthy population, particularly the wage earner.

PRIVATE ACTION

1868: The first major industrial medical care prepayment program, the hospital department of the Southern Pacific Railroad Co., was organized in Sacramento, California.

1872: The American Public Health Association (APHA) was organized. During its early years this organization was composed largely of administrative health officers who were concerned with public health in cities, states, and with the responsibilities of the Federal government in this field.

1877: Public health nursing was originated in the Women's Branch of the New York City Mission. The nurses were used in a program of home visiting and health supervision.

1879: The Mayo Clinic was founded in Rochester, Minnesota. It was a pioneer in the field of medical group practice in the United States.

1881: The American Red Cross was founded.

1882: The first major employee-sponsored mutual benefit association was the Northern Pacific Railway Beneficial Association which developed a program of complete medical care and other benefits financed by employer-employee payments.

PUBLIC ACTION

1869: The first State Board of Health in the United States was formed in Massachusetts.

1878: The National Quarantine Act, to prevent the introduction of contagious diseases into the United States, was passed.

1887: The first laboratory was established at the Marine Hospital on Staten Island. This "Laboratory of Hygiene" was reorganized, expanded, and became, as a result of a 1930 Act of Congress, the National Institute of Health.

1898: The United States Supreme Court made the first broad statement that the health of the laborer as a producer was considered to be as much a public benefit as the health of the consumer, and that the protection of labor becomes a public purpose.

1906: The Federal Food and Drug Act was passed. Its enforcement was begun in 1907.

1908: The first workmen's compensation law in the United States was enacted for the benefit of Federal Employees.

1887: Homestake Mining Co. of Lead, South Dakota, established a company-financed medical department with full-time staff which provided complete medical service to employees and their families.

1892: The Pennsylvania Society for the Prevention of Tuberculosis was founded. By 1904, other tuberculosis societies joined together to become the first of the national voluntary health agencies concerned with specific diseases or disabilities.

1904: The Council on Medical Education of the AMA was formed and began a program of accreditation of medical schools.

1909: The Metropolitan Life Insurance Co. began using Henry Street visiting nurses in an experiment in home nursing as a service to holders of industrial policies in a section of Manhattan in New York City.

1910: Health insurance plans which offered medical protection in the form of medical care for industrial workers in isolated areas, and disability benefits, were introduced by commercial and nonprofit organizations.

1910: The Flexner Report was published. This was a critical study of medical education in the United States and Canada which resulted in adoption of stringent measures to reduce total output of medical schools and improve the quality of their graduates.

1910: The first university degree in public health was awarded at the University of Michigan. A School of Public Health was jointly formed a few years later between the Massachusetts Institute of Technology and Harvard University.

1913: The first union health center was established by the International Ladies Garment Workers Union in New York City.

1911: Wisconsin became the first state to pass legislation that applied the workmen's compensation principles in an extensive fashion and survived the tests of the courts as to constitutionality. In the next decade, compensation laws were adopted in 42 states.

1912: The Children's Bureau was established in the Department of Labor by an Act of Congress. Among functions of this Bureau was the safeguarding of the health of mothers and children.

1920: The Vocational Rehabilitation Act passed by Congress was one of the first Federal grant-in-aid programs. It was originally conceived as a vocational training and counseling program for industrially injured civilians. Medical and physical restoration was not introduced as an important part of this program until 1943.

TABLE 3 (*cont.*)

PRIVATE ACTION (*cont.*)

1915: The first American Specialty Board was created to certify medical specialists in the field of ophthalmology.

1918: The American College of Surgeons developed a formal set of standards for hospitals including a requirement for regular review and analysis of clinical work of the hospital through medical staff conferences. This was a forerunner of hospital accreditation and of the medical audit.

1920: The American Medical Association made the first official declaration of its opposition to any compulsory scheme of health insurance controlled by any state or the Federal Government.

1920: The National Health Council was formed with an initial membership of 10 National Voluntary Health Agencies. The objective was to provide a mechanism for coordinating agency efforts.

1926: The first collective bargaining agreement with a health and welfare clause was negotiated between the Public Service Corp. of Newburgh, New York, and the Amalgamated Association of Street and Electric Employees.

1926: The Gies Report was published. It was a critical study of dental education in the United States and Canada, sponsored by the Carnegie Foundation for the Advancement of Teaching. It led to sweeping reforms in the field of dental education.

1927: The Committee on the Costs of Medical Care (CCMC) was "organized to study the economic aspects of the prevention and care of sickness, including the adequacy, availability, and compensation of the persons and agencies concerned." The Committee was financed entirely by voluntary and philanthropic contributions.

1929: The Baylor University Hospital Insurance plan was established by a contract between school teachers in Dallas, Texas and Baylor University Hospital; this was a forerunner of the Blue Cross hospital plans.

1929: The Community Hospital-Clinic of Elk City, Oklahoma was founded by the Farmer's Union Hospital Association. This was one of the first consumer cooperatives to provide medical care for members through group practice arrangements.

1929: The Ross-Loos Medical Group was founded in Los Angeles with the objective of providing low cost prepaid services to employees of the Los Angeles Water and Supply Department. This was one of the first of the private group clinics which offered prepaid services.

TABLE 4. A DECADE OF DROUGHT, DARKNESS, AND DEPRESSION
(1930-1939)

A period of the greatest prosperity the world had ever known ended abruptly in 1930 and was supplanted by worldwide depression. Recognition of the interdependence of the population was brought forth brutally by the physical and economic ravages which all states and communities shared. It was seen clearly that Federal responsibility for equalizing the burdens was a necessity if the nation was to prosper once again. One of the most significant ways this was evidenced was through passage of the Social Security Act.

PRIVATE ACTION

1931: The American Medical Association established its Bureau of Medical Economics indicating its growing interest in the economic problems of medical care. The purpose of the Bureau was to study the financing of medical care and suggest ways and means of improving existing systems of payment.

1933: The Committee on the Costs of Medical Care issued its final reports based on a five-year study of all types of medical services. The majority recommended that medical services could be organized to serve the entire population by the application of group practice units and insurance principles with financing either by private or governmental sources. The minority objected to any form of insurance covering physicians' services unless it was sponsored and controlled by medical societies.

1933: The American Hospital Association endorsed hospital prepayment insurance plans and established a list of essentials which should characterize such plans.

1933: The American Medical Association House of Delegates approved the Minority Report of the CCMC. This marked the beginning of the AMA's policy of encouraging voluntary insurance under medical control. It subsequently issued the following 10 principles as guides:

PUBLIC ACTION

1931: The Temporary Emergency Relief Act was developed and passed in New York State. Its provisions were the model for the Federal Emergency Relief Act passed in 1933.

1933: The Federal Emergency Relief Administration was created to assist states, through grants-in-aid, in maintaining their unemployment relief programs. This bill was expanded to provide medical attendance and medical supplies to recipients of unemployment relief programs. This program created public awareness of health needs and of the inadequacy of existing facilities especially in rural areas.

1934: New York was the first state to pass an enabling act permitting the establishment of nonprofit hospital service corporations under the State Insurance Commissioner.

1935-36: A national survey of the health of the population was conducted by the Public Health Service, which, with the CCMC, became the sole source of health data for the nation until 1950.

1935: The Social Security Act was passed. The legislation provided that the Federal government share in money payments to the needy aged, blind, and dependent children on public assistance. Title V authorized grants to states for maternal and child health and welfare programs which could include payment for direct personal care services. Title VI authorized annual appropriations for grants to states to assist in setting up and maintaining more adequate state and local health services, particularly public health services in rural areas.

1935: An Interdepartmental Committee was appointed by the President to coordinate health and welfare activities and to suggest improvements in the Social Security Program. Among its recommendations, outlining the Federal role in medical care, were the following:

1. Expansion of general physician's services
   a. In public health organizations and in combating specific diseases
   b. In maternal and child health services
2. Expansion of hospital facilities
3. Medical care of medically needy
4. A general program of medical care
5. Insurance against loss of wages during sickness

1936: The prepayment medical care program of the Farm Security Administration was implemented.

1938: Expansion of the PHS, construction of hospitals, and studies of methods, needs, and resources of public medical care for the indigent was proposed by the National Health Conference of 1938.

1938: The Railroad Retirement Act was passed and included provisions for sickness and maternity benefits.

1939: National compulsory health insurance for almost all employees and their dependents was proposed by Senator Wagner. Proposed benefits included physicians' services, hospitalization,

1. Medical sponsorship and control of plans;
2. No third party intrusion in physician-patient relationship;
3. Free choice of physicians;
4. Physician and patient relationship must be confidential;
5. Separate administration of hospital and medical plans;
6. Cost of medical care determined by income of patient;
7. Medical and disability benefits under separate administration;
8. Plans open to all qualified physicians on a voluntary basis;
9. Medical relief limited to those below "comfort level";
10. Rules and regulations governing medical care to be established by the medical profession only.

1934: The American College of Surgeons approved prepayment plans for medical and hospital service.

1934: Commercial insurance against the costs of hospitalization was first offered by private insurance companies.

1934: Formal principles were adopted by the American Dental Association for the dental aspects of health programs similar to the principles adopted by the American Medical Association. Its statement emphasized quality of care, methods of providing service, flexibility, free choice, and exclusion of non-professional profit seeking agencies.

1938: The Board of Trustees of the American Hospital Association established a Council on Hospital Care Insurance to coordinate the hospital insurance movement throughout the United States.

1938: The American Medical Association House of Delegates approved in principle tax-supported medicine for the indigent and voluntary health insurance for those above the level of indigency.

161

TABLE 4 (cont.)

PRIVATE ACTION (cont.)

1939: The first state-wide prepayment plans for physician services were organized by the state medical societies of California and Michigan. California Physician Services began operating the same year.

PUBLIC ACTION (cont.)

1939: National compulsory health insurance (cont.) drugs, and lab diagnostic services. Costs were to be covered through employer and employee contributions which were to have been deposited in a health insurance fund. This plan was to be administered through the states. The bill was not passed.

1939: The Federal Security Agency was formed and included the Office of Education, Public Health Service, Food and Drug Administration, Social Security Administration and, in 1943, the Office of Vocational Rehabilitation. In 1946, the Children's Bureau was added.

1939-42: There was increasing interest in compulsory disability insurance within the framework of the unemployment compensation systems, whereby a wage earner is indemnified in cash for unemployment due to non-occupational illness or injury sickness. Bills of this kind were introduced in 1939 in California, Massachusetts, New York, New Hampshire, and Pennsylvania. The first bill was enacted in 1942 by Rhode Island; California, New Jersey, and New York followed in subsequent years.

TABLE 5. WAR, PEACE, AND PROSPERITY
(1940-1969)

Under the stimulus of wartime needs and pressures and postwar prosperity, medical progress made significant strides on many fronts: penicillin and antibiotics; cardiac surgery; physical rehabilitation; and psychiatric therapy to name a few. The organization and financing of health services were also affected. During the war, the freezing of wages led to the increased acceptance of health benefits as an appropriate area for the collective bargaining process, thus stimulating the growth of prepaid health insurance. Impetus was also given to the development of needed health facilities and health manpower resources through shared public and private financing.

PRIVATE ACTION

1943: The first medical benefits, other than weekly sick benefits, provided under a collective bargaining agreement, became effective for members of the Philadelphia Waist and Dress Joint Board of International Ladies Garment Worker's Union.

1943: The American Medical Association created the Council on Medical Service and Public Relations to study the economic aspects of health insurance plans being introduced into the Congress.

1945: The Association of Medical Care Plans, with the Blue Shield as an emblem, was created as a national coordinating agency for physician sponsored health insurance plans.

1945: The Gunn-Platt Report, sponsored by the National Health Council, was published. This was an interpretive and critical study of the national voluntary health agency movement.

1946: The Health Insurance Council was formed by eight trade associations in the insurance field to serve as a liaison agency between the dispensers of health services, such as physicians and hospitals, and the underwriters.

1947: Medical services under the United Mine Workers Welfare and Retirement Fund were first provided to paraplegic miners.

PUBLIC ACTION

1940-45: Several bills were introduced into Congress to expand Social Security insurance benefits to cover hospital care for most of the working population and their families. None of these bills was enacted.

1943: The Emergency Maternity and Infant Care Program was established. This was administered by the Children's Bureau in the Department of Labor through Federal grants to state health departments. The program authorized maternity and infant care for wives and infants of enlisted men in specified grades of the Armed Forces, without cost and without a means test. It also stipulated that only those services which met minimum defined standards of care would be paid for.

1943: The United States Supreme Court upheld the verdict of two lower courts and found the American Medical Association and the Medical Society of the District of Columbia guilty of "restraint of trade" under the Sherman Anti-Trust Act as a result of organized medicine's opposition to the Group Health Association of Washington, D.C. GHA is a comprehensive prepaid group practice plan organized in 1937 as a consumer cooperative. This verdict set an important legal precedent denying the right of local medical societies to prevent the establishment or expansion of prepaid group practice.

163

TABLE 5 (*cont.*)

PUBLIC ACTION (*cont.*)

1946: The Hospital Survey and Construction Act authorizing the PHS to make grants to the states for surveying their needs for hospitals and public health centers, for planning construction of additional facilities, and assistance in financing such construction, was passed. The program is popularly known as Hill-Burton, after its two principal supporters.

1948: The National Labor Relations Board ruled that pension, health, and welfare plans were within the scope of collective bargaining. These rulings were sustained by action of the higher courts in 1949.

1948: The National Health Assembly was held under the sponsorship of the Federal Security Agency to formulate recommendations and guidelines for national policies in the health field.

1950: The Social Security Act was amended to provide Federal grants-in-aid to needy persons who are permanently and totally disabled. In addition, Federal matching funds were authorized to be used for direct payments for the medical care of public assistance recipients. This established the vendor payment principle.

1950: The first Presidential Commission on the Health Needs of the Nation was formed. Voluntary health insurance was recognized as a force throughout the nation.

1952: The first bill proposing to limit Social Security Health Insurance benefits to beneficiaries of Old Age Survivors and Disability Insurance was introduced in Congress by Senator Murray, Rep. Dingell, and others. Neither this bill nor others similar to it were passed.

PRIVATE ACTION (*cont.*)

1948: The American Public Health Association issued a statement on the "Quality of Medical Care and the Health of the Nation."

1948: The Medical Care Section of the American Public Health Association was formed.

1949: The Health Information Foundation was started by drug, pharmaceutical, chemical, and allied industries to conduct research in the social and economic aspects of the health field.

1949: The American Dental Association adopted a policy statement that voluntary prepayment and postpayment programs consistent with sound experience should be developed as rapidly as possible. It also issued criteria for evaluation of prepayment plans.

1949: As one of the recommendations of the National Health Assembly, the Commission on Chronic Illness was organized under sponsorship of the American Hospital Association, the American Public Health Association, and the American Public Welfare Association. The Commission conducted studies in the areas of the prevention of chronic illness, the care of long-term patients, and the prevalence and problems of chronic illness in a rural area and in an urban center. Its reports were published in four volumes in 1957.

1952: A joint statement by the American Public Health Association and the American Public Welfare Association was issued on the subject of "Tax-Supported Medical Care for the Needy."

1954: The Washington Dental Service Corporation, the first professionally sponsored nonprofit organization to provide prepaid dental care, was established.

1953: The Department of Health, Education and Welfare was created from the Federal Security Administration, giving cabinet status to these programs.

1954: Amendments to the Hill-Burton legislation expanded the program to include grants for construction of nursing homes, diagnostic and treatment centers, and rehabilitation facilities.

1954: Amendments to the Social Security Act provided for the "disability freeze." Under these, periods in which a worker has been totally disabled are omitted in computing his insured status and the average earnings on which his eventual benefits are based.

1956: The Social Security Act was amended to provide cash benefits as part of the old age and survivors insurance system to insured persons who are totally disabled for an extended period of time beginning at age 50. The amendments of this year also emphasized the need to help public assistance recipients toward independent living and to strengthen and maintain family life. The importance of rehabilitation was recognized by providing for the referral of the disabled to the state vocational rehabilitation agencies. The amendments also provided for Federal matching grants for medical care expenditures made in behalf of recipients of public assistance.

1956: The Congress passed the National Health Survey Act which authorized the Surgeon-General of the Public Health Service to conduct a continuing survey of illness and disability in the nation.

1960: The Kerr-Mills Act amended the Social Security Act to provide Federal-state matching grants for the establishment of programs for medical assistance to the aged. This was the first assistance program concerned entirely with vendor payments for medical care. The federal plan proposed liberal eligibility criteria,

1959: A statement of the American Public Health Association identified the role of public health in medical care as providing leadership in setting standards and promoting quality and efficient use of resources through health education, demonstration, and research.

1963: The APHA supported the principle that comprehensive medical care of good quality can and should be available to all persons in our society. The Association proposed that this can be facilitated through improving the organization of services within the framework of community health service centers under dynamic public health leadership.

1964: The American Dental Association House of Delegates approved the establishment of a national coordinating agency for professionally sponsored dental prepayment organizations.

1965: The Association of American Medical Colleges commissioned a study of its functions and structure, and particularly of its future role. "Planning for Medical Progress through Education," the completed report, recommended a broader role for the Association in furthering medical education.

1966: The Citizens' Commission on Graduate Medical Education, created by the American Medical Association, issued a first report, "The Graduate Education of Physicians." Among the major recommendations: abandonment of the traditional year of internship and creation of a new specialty composed of "primary physicians."

1966: The American Medical Association's Council on Medical Service and Judicial Council issued a joint statement on financing medical care stating that the medical profession cannot dictate to patients how they shall finance their medical bills.

TABLE 5 (cont.)

PRIVATE ACTION (cont.)

1966: The National Commission on Community Health Services submitted its report to the President. It identified 13 major health problems and made over 100 recommendations for improving health services. The report resulted from a 4-year study by the private commission which was sponsored by the American Public Health Association and the National Health Council.

1967: An Advisory Commission on Health Care of the American People was established by the American Medical Association. The membership is composed of nationally known representatives from a broad range of activities.

1967: The American Medical Association sponsored the First National Conference on the Socioeconomics of Health Care. Areas covered were the impact of medical and social changes on health care, the role of the hospital and medical staff in the community, and new resources and methods in training and utilization of health manpower.

1967: The American Public Health Association's Governing Council adopted a policy statement in support of health planning in all its aspects and commended its use. Eight principles were issued to assist communities and agencies. A resolution concerning the rising costs of health care was also adopted, urging a Joint Congressional Committee to determine the facts and make recommendations for the guidance of both the public and private sectors of the health care economy.

1967: The 1967 National Health Forum, sponsored by the National Health Council, was devoted to planning for health. Leaders in the field of health and medicine attempted to shape some general guidelines for comprehensive planning in the states and communities.

PUBLIC ACTION (cont.)

1960: The Kerr-Mills Act (cont.) comprehensive medical benefits, and up to 80 percent Federal payments under open-end appropriations. It also expanded the medical care program of old age assistance.

1960: The Federal Employees Health Benefits Program, in which the Federal government shares with the employee the cost of health insurance for himself and his family purchased from specified private carriers, was enacted.

1961-62: The 87th Congress introduced at least 20 bills to finance some portion of the aged's medical care costs. None passed.

1961: The Community Health Services and Facilities Act was passed. It provided formula grants to states for improving health services to chronically ill and aged and project grants to improve methods for providing health services outside of hospitals.

1961: The Health Services Research Study Section of the PHS was formed as a result of the final report of the Surgeon-General's study group on the Mission and Organization of the PHS. A part of every PHS program was to undertake research, not only intramurally, but also extramurally.

1961: The insurance Commissioner of the Commonwealth of Pennsylvania issued an adjudication which emphasized the public nature of such voluntary health insurance plans as Blue Cross and Blue Shield, and required that they have a greater degree of public accountability.

1964: The Health Professions Educational Assistance Act, providing grants for the construction of teaching facilities for medical,

1968: The American Medical Association sponsored national conferences during the year on the topics: "Community and Emergency Medical Services"; "Health Care of the Poor"; and "Utilization Review."

1968: The topic of the National Health Forum was "Quality of Health Services," based on the premise that responsibility for quality in health care should rest with the professionals who render this care, while recognizing that the consumer's satisfaction is the final measure.

dental, and other health personnel, was enacted. In addition, this Act provided funds for a student loan program and its primary purpose was to increase the nation's supply of professional health personnel. A program to provide assistance for training professional nurses was also enacted.

1964: The Legislature of the State of New York enacted a Bill (The Metcalf Bill) giving the State final determination as to the actual need for proposed new health facilities as well as for substantial modification of existing facilities. This was the first legislation which gave the government an effective role in controlling hospital and related facility planning.

1964: The President's Commission on Heart Disease, Cancer and Stroke completed its study and reported 35 major recommendations for a national program to conquer these major diseases.

1965: The Congress amended the Social Security Act to: (1) establish a new Title XVIII providing for a program of health insurance for the aged, including a hospital insurance program financed through social security taxes and a voluntary supplementary medical benefits program financed by matching contributions from beneficiaries and from the general revenues; (2) establish a new Title XIX providing for an expanded and unified program of grants to states for medical assistance which could eventually replace existing medical care provisions under all public assistance programs; (3) increase appropriations and establish new grant programs related to maternal and child health and crippled children's services; (4) increase social security cash benefits; and (5) improve the Federal-state public assistance programs.

1965: The Appalachian Regional Development Act (PL 89-4) was passed, providing construction grants for demonstration health facilities in the Appalachian region and additional support for initial operating costs.

TABLE 5 (*cont.*)

PUBLIC ACTION (*cont.*)

1965: The Health Professions Education Act (PL 89-290) made grants available to nonprofit schools of medicine, dentistry, osteopathy, optometry, and podiatry for improving their educational programs. It also began scholarship grants for schools to award to students who could not otherwise afford specialized study in the health professions.

1965: The Regional Medical Programs (PL 89-239) established grants to plan and develop regional cooperative arrangements by public and nonprofit groups for demonstration, training, and research in patient care in the field of heart disease, cancer, stroke, and related diseases.

1965: The Community Mental Health Centers Act (PL 89-105) authorized grants to public or nonprofit private organizations to help pay for professional and technical staff in community mental health centers.

1965: The White House Conference on Health was called to bring together "the best minds and the boldest ideas to deal with the pressing health needs of the nation." The focus of discussion was on the poor and minority groups as forgotten patients, emphasizing the need to expand health care.

1966: The Allied Health Professions Training & Construction Act (PL 89-751) was enacted to improve the training of personnel in the allied health professions by providing grants for construction, improvement, traineeships, and development of new methods.

1966: The Comprehensive Health Planning & Services Act (PL 89-749) introduced the concept of comprehensive health planning as a mechanism through which the planning activities of all the elements involved in or concerned with health services can be linked together. It enables governmental and nongovernmental health and related agencies and groups to develop a cooperative and coordinated approach toward the objective of good health for every individual. It also gives states greater flexibility in supporting health services in the communities. Federal funds for state public health programs were formerly categorized for specific disease problems. This legislation removed these categorical restrictions, allowing the states to channel grant funds into those health services which the states determine are most needed. Major program provisions: (a) Formula grants to states for comprehensive health planning, each state designating a single state agency to administer the planning process. (b) Public or nonprofit private agencies or organizations may apply for grants to develop areawide planning. These must be approved by the state agency to assure coordination. (c) Provides project grants for training studies and demonstrations in health planning to public or nonprofit private agencies, institutions, or organizations, including universities. (d) Grants to states to establish and maintain adequate public health services of all kinds, including training of personnel for state and local health work. States are required to submit a state plan for provision of public health and mental health services.

1966: The Economic Opportunity Amendments of 1966 (PL 89-794), through Community Action Programs, authorized the Office of Economic Opportunity to make grants to or contract with public or private nonprofit agencies to operate neighborhood health centers in both urban and rural areas of concentrated poverty. These centers provide a "one door" approach to comprehensive

health services, provide for training of indigenous native persons in new health careers, and use consumer boards to advise on program development and operations.

1966: The Demonstration Cities and Metropolitan Development Act (PL 89-754) contained a program of FHA mortgage insurance to finance construction and equip facilities for group medical, dental, and optometric practice.

1966: A Medical Care Costs study issued by the Department of Health, Education and Welfare, found no evidence that the Medicare program was causing price rises. It said the demand for physician services had outstripped the supply and physicians had responded by both increasing their productivity and raising their prices.

1967: The National Conference on Medical Costs, which followed the Federal Study of Medical Care Costs, found that the health care industry is uncoordinated. It noted that health insurance had contributed to this fragmentation of medical services. Alternatives to the present organization, such as group practice of medicine, were suggested.

1967: The National Conference on Private Health Insurance criticized the carriers for tending to insure only hospitalization, the most expensive form of medical care. The sponsoring Department of Health, Education and Welfare placed the responsibility on the insurance companies for improving the financing of health care.

1967: A National Conference on Group Practice noted that physicians are recognizing the advantages of group practice: the

TABLE 5 (*cont.*)

PUBLIC ACTION (*cont.*)

1967 (*cont.*)
number of groups nearly tripled between 1959 and 1966 and the number of physicians in group practice had nearly doubled. However, comprehensive prepaid plans had failed to keep pace.

1967: The report of the National Advisory Commission on Health Manpower, appointed by the President to seek ways to ease a "critical shortage of qualified health personnel," recommended training more physicians and other health workers, providing incentives for efficient operation of hospitals, and improving the quality of physicians' services through "peer group" review.

1967: The Social Security Amendments of 1967 modified the provisions of the Medicare and Medicaid legislation to, among other things: (a) simplify certain payment procedures, (b) consolidate all outpatient coverage under the supplementary medical insurance program, (c) pay charges of hospital-based physicians, (d) expand certain benefits such as in-hospital days and outpatient physical therapy, (e) increase the premium rate for medical insurance, and (f) allow the Secretary of HEW to experiment with alternative

1967 (*cont.*)
methods of reimbursement in order to test incentives for increased efficiency.

1967: The Partnership for Health Amendments (PL 90-247) initiated Federal licensing of clinical laboratories dealing in interstate commerce, and established a program of grants for research and demonstrations to develop more efficient health services and facilities.

1968: The Department of Health, Education and Welfare Task Force on prescription drugs, was formed after a number of legislative attempts to include drugs at generic prices as part of Federal health programs. The Task Force suggested ways to improve the quality of drugs and to control costs.

1968: The report of the President's National Advisory Commission on Health Facilities stated that the development of systems for the delivery of comprehensive health care to all our people where they live is a community responsibility. Coordination of resources and services must be made at the local level, with professional direction.

Appendix D

RESOURCES FOR COMMUNITY HEALTH DATA

U.S. DHEW, HSMHA, National Center for Health Statistics. Vital and Health Statistics Reports. U.S. Government Printing Office, Washington, D.C.

U.S. DHEW, HSMHA, National Center for Health Statistics. Monthly Vital Statistics Report. U.S. Government Printing Office, Washington, D.C.

U.S. DHEW, HSMHA, Center for Disease Control. Morbidity and Mortality Weekly Report. U.S. Government Printing Office, Washington, D.C.

U.S. DHEW, HSMHA, Community Health Service. The Community Profile Data Center, Parklawn Building, Rockville, Md.

U.S. Dept. of Commerce. Bureau of the Census. Census Use Study: Health Information System II. Report No. 12. Washington, D.C.

County and City Data Book, Census Tract Series. Family Health Survey.

U.S. DHEW, PHS, Bureau of Community Environment Management. "NEEDS," Neighborhood Environmental Evaluation and Decision System. U.S. Government Printing Office, Washington, D.C.

State and Local Office of Vital Statistics

Systems Development Corporation. A Geographic Base File For Urban Data Systems. Santa Monica, California.

Appendix E
DESCRIBING EVENTS NUMERICALLY

DESCRIPTIVE STATISTICS

Measures of Central Tendency

$$\text{Mean} = \overline{X} = \frac{\Sigma X}{N} = \frac{\text{sum of observations}}{\text{number of observations}}$$

most frequently used measure of central tendency
Median = mdn = midpoint in the distribution
Mode = most commonly occurring observation or score
Shape of distribution—is the shape symmetrical (bell-shaped curve) or skewed

Meaures of Variablity

Range = spread of observations or scores = difference between the highest and the lowest.
Standard deviation = SD = σ = $\sqrt{\dfrac{\Sigma (X-\overline{X})^2}{N}}$ =

relationship of observations or scores to the mean.

Correlation

Coefficient of correlation = the degree to which scores fall along the regression line = a measurement of relationship
Factor analysis = the manner in which a number of variables (factors) are related.

Vital Statistics (Demography)

Rate = the rapidity with which an event is occurring

$$\text{Birth rate} = \frac{\text{number of live births during the calendar year}}{\text{population at midyear}} \times 1{,}000$$

$$\text{Crude death rate} = \frac{\text{deaths among residents during the calendar year}}{\text{population at midyear}} \times 1{,}000$$

$$\text{Prevalence rate} = \frac{\text{number of known cases at a specific point in a time period}}{\text{population at that time}} \times \begin{matrix} 100 \\ 1{,}000 \\ 10{,}000 \\ 100{,}000 \end{matrix}$$

$$\text{Incidence rate} = \frac{\text{number of new cases in defined time period}}{\text{population at midpoint}} \times \begin{matrix} 100 \\ 1{,}000 \\ 100{,}000 \\ 1{,}000{,}000 \end{matrix}$$

INFERENTIAL STATISTICS

Sample = a small portion of a defined population deemed to be representative of the total

Standard error of the mean = SE_X + standard deviation of the many sample means = $\dfrac{SD}{\sqrt{N}}$

levels of confidence = numerical standard error of the mean that will be tolerated

t-test = a test developed to evaluate the difference between small groups

analysis of variance = a test developed to compare simultaneously the scores of several groups

Chi square = χ^2 = a nonparametric statistic designed to test relationship between two variables

probability level = p = the difference in proportions did not happen by chance.